MW01516780

The Neuroethics of Memory

The Neuroethics of Memory is a thematically integrated analysis and discussion of neuroethical questions about memory capacity and content, as well as interventions to alter it. These include: How does memory function enable agency, and how does memory dysfunction disable it? To what extent is identity based on our capacity to accurately recall the past? Could a person who becomes aware during surgery be harmed if they have no memory of the experience? How do we weigh the benefits and risks of brain implants designed to enhance, weaken or erase memory? Can a person be responsible for an action if they do not recall it? Would a victim of an assault have an obligation to retain a memory of this act, or the right to erase it? This book uses a framework informed by neuroscience, psychology and philosophy combined with actual and hypothetical cases to examine these and related questions.

WALTER GLANNON is Professor of Philosophy at the University of Calgary, Canada. He has served as a clinical ethicist at three different hospitals and has held academic appointments at McGill University and the University of British Columbia. He is the author or editor of ten books, including *Free Will and the Brain: Neuroscientific, Philosophical and Legal Perspectives* (Cambridge University Press, 2015).

The Neuroethics of Memory

From Total Recall to Oblivion

Walter Glannon

CAMBRIDGE
UNIVERSITY PRESS

CAMBRIDGE
UNIVERSITY PRESS

University Printing House, Cambridge CB2 8BS, United Kingdom

One Liberty Plaza, 20th Floor, New York, NY 10006, USA

477 Williamstown Road, Port Melbourne, VIC 3207, Australia

314–321, 3rd Floor, Plot 3, Splendor Forum, Jasola District Centre, New Delhi – 110025, India

79 Anson Road, #06-04/06, Singapore 079906

Cambridge University Press is part of the University of Cambridge.

It furthers the University's mission by disseminating knowledge in the pursuit of education, learning, and research at the highest international levels of excellence.

www.cambridge.org
Information on this title: www.cambridge.org/9781107131972
DOI: 10.1017/9781316443712

© Walter Glannon 2019

This publication is in copyright. Subject to statutory exception and to the provisions of relevant collective licensing agreements, no reproduction of any part may take place without the written permission of Cambridge University Press.

First published 2019

Printed in the United Kingdom by TJ International Ltd, Padstow Cornwall

A catalogue record for this publication is available from the British Library.

Library of Congress Cataloging-in-Publication Data
Names: Glannon, Walter, author.
Title: The neuroethics of memory : from total recall to oblivion /
    Walter Glannon, University of Calgary.
Description: New York : Cambridge University Press, 2019. |
    Includes bibliographical references and index.
Identifiers: LCCN 2019009277| ISBN 9781107131972 (hardback : alk.
    paper) | ISBN 9781107583412 (pbk. : alk. paper)
Subjects: LCSH: Neurosciences–Moral and ethical aspects. | Memory–Moral
    and ethical aspects. | Medical ethics.
Classification: LCC RC343 .G535 2019 | DDC 612.8/233–dc23
LC record available at https://lccn.loc.gov/2019009277

ISBN 978-1-107-13197-2 Hardback
ISBN 978-1-107-58341-2 Paperback

Cambridge University Press has no responsibility for the persistence or accuracy of URLs for external or third-party internet websites referred to in this publication and does not guarantee that any content on such websites is, or will remain, accurate or appropriate.

For Yee-Wah

Contents

List of Figures		*page* viii
Acknowledgments		ix
	Introduction	1
1	Memory Systems and Memory Stages	14
2	Agency, Identity and Dementia	50
3	Anesthesia, Amnesia and Recall	84
4	Disorders of Memory Content and Interventions	112
5	Disorders of Memory Capacity and Interventions	140
6	Legal Issues Involving Memory	169
	Epilogue: The Future of Memory	196
	References	202
	Index	225

Figures

1.1 Taxonomy of memory (with permission from
    Larry Squire)   *page* 31
1.2 Mapping memory (with permission from
    Christian Ineichen)   32
1.3 Memory consolidation and reconsolidation
    (with permission from Karen Rommelfanger)   36
1.4 Brain activation in episodic memory retrieval
    (with permission from Nick Davis)   38
5.1 Memory prosthesis (with permission from
    Theodore Berger)   157

Acknowledgments

I thank my editors at Cambridge University Press, Hetty Marx, Janka Romero, Emily Watton and Bethany Johnson for their advice and support. Three reviewers commissioned by the Press gave me constructive comments and suggestions on the book proposal. One of the reviewers, Nick Davis, gave me especially helpful comments on the entire manuscript. Some of the material in the book is based on previously published journal articles. I am grateful to the referees of these articles for their thoughtful reviews.

I presented some of the ideas in the book to audiences at conferences and workshops in Geneva, Halifax, Lausanne, Montreal and Muenster. For correspondence and discussion of some of the ideas in the book, I thank Jackie Andrade, Jared Craig, Andrew Davidson, Paulie Gaido, Christian Ineichen, Kate Johnson, Adam Kolber, Timothy Krahn, Jaakko Langsjo, Neil Levy, Ying-Tung Lin, Stephen Morse, Roger Pitman, Dimitris Repantis, Ulrike Rimmele, Karen Rommelfanger, Julian Savulescu, Daniel Schacter, Dominic Wilkinson, Karsten Witt, Teresa Yu and Adam Zeman.

.

Introduction

In classical mythology, the souls of the dead who drank from the river Mnemosyne would remember everything from their experience. Those who drank from the river Lethe would forget everything and enter the realm of oblivion. In our actual lives, memory capacity and impairment fall along a spectrum between these two extremes. Unlike the characters in mythology, we cannot choose how much or how little memory we have. Working memory, retrieval of episodic and semantic memory and the initial learning in procedural memory are to some extent within our conscious control. But we have no control over the encoding, consolidation, storage and reconsolidation of memory. This may change, however, with interventions designed to increase memory capacity, alter the content of memories or erase them.

Memory is a vital process in humans. At the most basic biological level, the capacity of the adaptive arm of the immune system to form a memory of antigens enables it to recognize and eliminate pathogens through the combined action of antibodies, complement and macrophages. Antigenic memory is thus necessary for the survival of the organism. At more evolved neurobiological and psychological levels, learning mediated by subcortical brain structures enables us to perform motor skills automatically without having to think about performing them. Brainstem structures responding to sensory stimuli send inputs to the hippocampus that allow us to remember new places. Memories of threatening events mediated by the brain's fear memory system allow us to recognize new threats and confront or avoid them. These are further examples of how memory is critical to our survival. At a psychological level, the experience of mental time travel in recalling the past and imagining the future gives one the feeling of persisting through time as the same person. It allows one to integrate one's experiences into a coherent whole and construct a meaningful autobiography. Information about the past enables us to engage in goal-directed behavior in forming and executing action plans.

Robert Veselis emphasizes the significance of the memory process: "Memory makes us uniquely human. As the human mind is the most complex creation in the universe, it stands to reason that memory embodies to a large extent this complexity. When memory fails in the end for some of us, a large portion of our being human also fails" (Veselis, 2017, p. 31).

Visual, auditory, gustatory and olfactory cues can trigger autobiographical memories transporting us back to childhood or places where we experienced certain sights, sounds, tastes and smells. Memories of the departed allow them to visit us in dreams. They may console or haunt us. The Ghost in *Hamlet* is the mental representation of the main character's father. In the same play, the graveyard scene in which Hamlet reflects on the deceased court jester Yorick when his skull is exhumed is another example of the power of episodic memory. Recall of past misdeeds or omissions can generate regret and other emotions that can influence our current and future behavior in beneficial or harmful ways. Memories others have of us may provide a sense of virtual survival beyond death. But all of these memories eventually dissolve in oblivion.

Brain injury and neurological disorders can disrupt the brain's capacity to encode, consolidate, store, retrieve and reconsolidate memories. This disruption can adversely affect the psychological capacities associated with memory and have a deleterious effect on people's lives. The inability to form new memories or retrieve existing memories can impair or undermine the experience of persisting through time and the capacity for agency. In other circumstances, an emotionally charged memory of a traumatic experience may become firmly entrenched in the brain and mind and cause the psychopathology characteristic of posttraumatic stress disorder (PTSD), panic and anxiety disorders. Depending on how it affects our thought and behavior, memory can have value or disvalue for us.

This book is a thematically integrated analysis and discussion of neuroethical questions about memory and interventions to modify it. It is written for a multidisciplinary audience, including psychologists, clinical neuroscientists, philosophers, bioethicists, legal theorists and informed lay readers. By discussing historical and current theories of memory, and examining existing and emerging forms of memory modification, the book shows how empirical and normative aspects of memory have evolved. The subtitle of the book captures the spectrum of memory extending from exceptional recall to profound amnesia and advanced dementia. These opposite ends of the spectrum are the rough equivalents of the mythological Mnemosyne and Lethe. Our ability to adapt to changing environments requires optimal levels of memory capacity and content between these extremes. A certain amount of information about the past is necessary to plan and act in the present and future. But too much of this information can overload the brain and mind and interfere with reasoning and decision-making. Some degree of forgetting is necessary to learn new information relevant to one's natural and social milieu. Flexible thought and behavior require a balance between remembering and forgetting.

Although different memory systems support different physical and mental functions, I focus mainly on two subtypes of declarative memory: episodic and semantic. Episodic memory is knowledge of events that happened at a specific time and place (Tulving, 1983). Semantic memory is knowledge of facts and concepts about the world (Tulving, 1985a). Working and prospective memory are necessary for rational and moral agency. These two declarative memory subtypes rely on episodic and semantic memory, which have a broader range of functions. Episodic memory is necessary not only for agency but also for identity. Among memory systems, episodic and semantic memory are most pertinent to metaphysical, ethical and legal questions about identity, agency, responsibility, benefit and harm.

More specifically, how do normal memory functions enable us to initiate and execute action plans? How does memory dysfunction impair this ability? To what extent is personal identity based on the capacity to accurately recall the past? How many memories could be lost without causing a substantial change in identity? Could a person with early-stage dementia exercise precedent autonomy in expressing earlier wishes about later life-sustaining care when she has no memory of these wishes? If a patient under general anesthesia becomes aware intraoperatively, then would it be permissible for an anesthetist to infuse an amnesia-inducing drug without the patient's prior consent? Would the patient be harmed if she had no memory of being aware? How would the patient know that she was aware without a memory of it? How do we weigh the potential neurological and psychological benefit against the risk of harm from brain implants designed to improve or restore some memories and weaken or erase others? Do we discover our true selves through the backward-looking aspects of memory or create them through the forward-looking aspects of memory? How would modifying memories influence authenticity? Can a person be responsible for an action if she does not remember performing it? Would a victim of a criminal act have to duty to retain a memory of it to testify against the perpetrator? Or would her cognitive liberty give her the right to erase the memory? Focusing mainly on disorders of memory content and capacity, I use actual and hypothetical cases to analyze and discuss these questions.

What it is like to recall an experience is more than a function of neurobiology. Still, we cannot understand memory without understanding its neurobiological underpinning. Interactions between cortical and subcortical brain regions allow the integration of information about a person's experience into a coherent and consciously accessible representation of it. Although our experience as rational and moral agents and subjects persisting through time is a psychological property, it is possible through the normal function of neurobiological processes that enable memory. Neuroscientific and behavioral research has helped to explain normal memory function, as well as how transient and chronic neurological disorders can result in different types of memory dysfunction. This research on memory and how it influences our thought and behavior forms the theoretical basis of this book.

While I explore the philosophical *implications* of memory, I do not engage with philosophical *theories* of memory (Sutton, 1998; Bernecker, 2008, 2010; Bernecker and Michaelian, 2017). Many of these theories do not adequately consider the multiple neural networks that mediate different memory functions. Some ignore these networks altogether and discuss memory exclusively in psychological terms. Yet failing to account for both psychological and neurobiological aspects of memory results in what is at best an incomplete explanation of the different systems, types and subtypes of memory. In normal circumstances, whether a person remembers events or facts can be confirmed by verbal reports and her general behavior. Yet memory disorders have provided the best evidence of memory and its role in our conscious and unconscious life. These usually result from brain damage and dysfunction. Some forms of amnesia may be psychogenic, though these tend to be transient. Dissociative disorders are typically described as psychogenic. But they may correlate with detectable neurobiological changes. Although they manifest in varying types and degrees of mental impairment, most memory disorders are associated with anatomical and functional abnormalities in the brain.

Some philosophers distinguish between experiential, propositional (or factual) and practical memory (Bernecker, 2008, 2010). This taxonomy is consistent with psychologists' and cognitive neuroscientists' distinction between episodic, semantic and procedural memory (Tulving, 1983, 1985a). Philosophers tend to explain memory content in terms of experiential (nonpropositional) or factual (propositional) attitudes. The first type of attitude has a first-person content, and the second type of attitude has a third-person content (Burge, 2003). These attitudes correspond roughly to autobiographical and semantic memory. Yet the abstract and at times overly technical formulation of them fails to show how they manifest in actual thought and behavior.

Distinguishing between direct realist and representational theories of memory, some philosophers discuss whether we have direct or indirect access to past events (Sutton 1998; Bernecker, 2008, 2010). According to direct realism, when we recall an event, we are in direct cognitive contact with it. According to representationalism, when we recall an event, we are aware of an imperfect mental idea of it that falls short of direct cognitive contact. Causal intermediaries between the experience and its recall may alter the content of the memory. The cognitive neuroscience of memory endorses some version of representationalism. Still, the idea we have of the experience is not just a mental process but also a neural process involving varying degrees of integrated information in the brain. The key causal intermediaries influencing the extent to which the representation resembles the experience are changing contextual factors in the time between the experience and the initial memory of it and later retrieval and reconsolidation of the memory.

Those who accept direct realism or representationalism tend not to fully appreciate the concept of memory as a dynamic process of continuous updating of information. They fail to appreciate the extent to which we consciously and unconsciously edit memory in our mental life. They focus too much on the "pastness" of memory and not enough on its future-oriented aspect. It is important to point out, though, that not all philosophers focus primarily on the past in discussing memory. Influenced by contemporary psychologists investigating memory, an increasing number of philosophers focus on the constructive aspect of memory and its neural correlates (e.g., De Brigard, 2013, 2017; Michaelian, 2016, pp. 82–85).

I do not engage with the literature on the politics of collective memory either (Blustein, 2008; Campbell, 2014; Stone and Bietti, 2016; Belavusau and Gliszczynska-Grabias, 2018). While memory of historical injustice toward certain groups generates an obligation for societies to restore justice, this type of memory is very different from the type I examine in this book. My concern is not with how groups remember the past but with how individuals remember it. I discuss memory as a neurobiological and psychological process rather than a social and political one. The general focus is on how information as a preserved neural and mental representation of the past shapes how we think and act in the present and future.

Normative questions about memory fall within the domain of neuroethics. This is an interdisciplinary field at the intersection of the clinical neurosciences, cognitive science, psychology, radiology (neuroimaging), philosophy and law. In a seminal paper published in 2002, Adina Roskies distinguished between two branches of neuroethics: the ethics of neuroscience and the neuroscience of ethics (Roskies, 2002). The first branch considers the potential benefits and risks to patients and research subjects whose brains are mapped or monitored by structural and functional imaging. It also considers the potential benefits and risks of altering the brain with psychotropic drugs, surgery and electrical stimulation. The ethics of neuroscience also considers the obligations of clinicians and investigators to protect patients and research subjects with neuropsychiatric disorders from harm. The neuroscience of ethics generally pertains to the neurobiological basis of rational and moral decision-making. It concerns the cognitive and emotional capacity to consider reasons for and against brain interventions, how one may be affected by them and how one may make informed decisions to receive or refuse treatment and participate or decline to participate in research. While acknowledging that the ethics of neuroscience and the neuroscience of ethics "can be pursued independently to a large extent," Roskies noted that "perhaps most intriguing is to contemplate how progress in each will affect the other" (2002, p. 21).

An example of research on a disorder of memory capacity illustrates how the two branches of neuroethics can overlap. Suppose that a researcher conducting a clinical trial on deep brain stimulation of the hippocampal-entorhinal circuit as a potential therapy for anterograde amnesia recruits a patient with this disorder. The researcher is obligated to obtain informed consent from the research subject and ensure that he is not exposed to more than minimal risk from the intervention (ethics of neuroscience). Although memory systems often interact, they are dissociable. While the hippocampal damage impairs the subject's ability to form new memories, the prefrontal region mediating executive functions associated with the subject's working memory may be intact. Other neocortical regions storing episodic and semantic information used in working memory may be intact as well. This may provide the subject with enough cognitive and emotional capacity to consider the prudential and moral reasons for participating in research. He may be able to consciously hold information long enough and exercise a sufficient degree of decisional capacity to give informed consent to participate in the trial. Yet if he has a dysfunctional prefrontal cortex and his working memory is impaired, then he may lack the cognitive capacity to make decisions and give informed consent (neuroscience of ethics). This is one of the challenges in recruiting early-stage Alzheimer's patients for research on techniques designed to improve memory function. Changes in their brains affecting working memory may impair the degree of decisional capacity necessary to consent to participate as subjects in clinical trials.

A different example of research on a disorder of memory content further illustrates the overlap between the ethics of neuroscience and the neuroscience of ethics. Suppose that a researcher conducting a clinical trial using a protein synthesis inhibitor to erase a pathological fear memory in the amygdala recruits a patient with PTSD as a research subject. Again, the researcher is obligated to protect the subject from harm by obtaining informed consent from her and ensuring that she is not exposed to more than minimal risk (ethics of neuroscience). The subject may have enough prefrontal-mediated cognitive function to process information about the trial, consider how she might contribute to and be affected by the trial and give informed consent to participate in it. However, if hyperactivation of her fear memory system projects to and impairs prefrontal function, then her cognitive capacity to weigh the potential benefits and risks of participating in the research could be impaired (neuroscience of ethics). Her hyperactive emotional state could compromise her capacity for informed consent. How one assesses questions in the neuroscience of ethics about whether a person affected by a memory disorder has the requisite decisional capacity may depend on the severity of the disorder. This capacity can be measured by combining neuroimaging or electrophysiological recording with assessment of the person's behavior.

The neuroethics of memory can be construed broadly to include not only questions about the potential benefit and harm of memory-modifying interventions in the brain. It also includes the effects of memory disorders and treatments for them on agency, identity and the role of memory in judgments of moral and criminal responsibility. Memory is thus relevant to issues in metaphysics, ethics and criminal law. These issues are informed by neurobiological and psychological determinations of memory function and dysfunction. An interdisciplinary perspective on memory corresponds to the interdisciplinary nature of neuroethics.

My discussion of the ethical and legal dimensions of memory is not driven by a single theory. Instead of selecting a theory and then discussing issues around it, I first raise the issues and then explicitly or implicitly apply a theory to explain why a course of action or policy is justified or unjustified. For example, whether memory should be enhanced or erased drives the application of an ethical theory to answer these questions. Because improvement in the capacity to consolidate and retrieve memories is the intended outcome of drugs or neurostimulation for disorders of memory capacity, the relevant ethical theory to assess these interventions is consequentialism. This same theory can provide a normative framework for addressing questions about the justification of erasing memories in disorders of memory content. In criminal law, the question of whether a victim of a crime with a traumatic memory of it has the right to erase the memory or is obligated to retain it to testify against the perpetrator involves balancing deontological and consequentialist considerations. One must balance the victim's right to eradicate the source of continuous harm against the public's interest in avoiding potential future harm from the offender. Both deontological and consequentialist considerations may also be relevant to questions of whether a person who committed a crime but cannot recall his action should be held responsible or punished for it.

Methodologically and structurally, the book is divided roughly into two parts. I first outline some of the history behind theories of memory and describe different memory systems and mechanisms of encoding, consolidating, storing, retrieving, reconsolidating and re-storing memories. These systems and mechanisms constitute the neurobiological and psychological framework of memory. Within this framework, I examine the role of normal and abnormal memory in human thought and behavior and some of the philosophical and legal issues that arise from measuring and modifying it. Many of these issues involve disorders of memory content and memory capacity. Both types of disorder involve dysfunction at earlier and later stages of the memory process. Although their effects can be different, they can be equally mentally disabling.

In Chapter 1, I trace the main historical developments in theories of memory from Aristotle's concept of recollection to the current model of episodic memory as a constructive and reconstructive process. This is the legacy of Frederic Bartlett's early twentieth-century concept of memory as a reconstruction rather than reproduction of the past. Then I describe the taxonomy of memory systems. I also describe memory as a process that extends from encoding and consolidation in earlier stages to retrieval and reconsolidation in later stages. I explain how different neural circuits regulate different stages in this process. Retrieval and reconsolidation are critical to reconstructing episodic memories because they enable us to update memories to make them relevant to our present and future circumstances. The interconnection between retrieval and reconsolidation in destabilizing and restabilizing memories is necessary to update them. In addition to making memory adaptive, retrieval is the stage of the memory process where altering the content of memories, or erasing them, may be possible because they are labile and susceptible to change at this stage.

I discuss the role of memory in agency and identity in Chapter 2. Working memory is necessary for executive functions in reasoning and decision-making. It does not operate alone but relies on stored episodic and semantic memory. Because agency also involves goal-directed behavior and future planning, prospective memory is also critical for agency in holding intentions over time. Impairment in any of these types of memory can interfere with the capacity to form and translate intentions into actions. Agency illustrates how different memory systems interact to enable the cognitive and emotional functions necessary for flexible thought and behavior. In addition, I discuss the role of episodic memory in the psychological connectedness and continuity that constitute personal identity. I describe how anterograde and retrograde amnesia can disrupt these psychological relations and thus identity. We constantly change the content of our memories by updating them. If the purpose of memory is not to reproduce the past but to enable us to simulate events in adapting to current and future environments, then only a critical core set of representations of the past may be necessary to provide a basis for these mental acts.

I explore the implications of radical life extension for memory and what it would mean for identity and self-regarding concern about the future. Specifically, I consider whether a person with a substantially longer life would become a different person with different interests beyond a certain point because of changes in the content of her episodic memories. If the adaptive function of memory involves changing its content and weakening of psychological connectedness and continuity over time, then adaptability may come at the expense of personal identity in a radically extended lifespan. I also discuss the loss of memory in dementia and the concept of precedent autonomy. This involves the question of whether a competent patient's wishes about life-sustaining or life-ending interventions apply when she is demented and has no memory of these wishes. If one accepts precedent autonomy, then the moral and legal force of a request made by a competent person in an advance directive transfers from the earlier to the later time.

In Chapter 3, I examine empirical, epistemological and ethical issues surrounding anesthesia awareness with postoperative recall. Some patients unexpectedly become aware during surgery despite receiving general anesthesia. If awareness cannot be detected intraoperatively, then a report from the patient may be the only way to confirm awareness. Yet a report of the experience requires a memory of it, and not all patients recall being aware. This raises the question of whether a patient could be harmed by becoming aware if she did not remember the experience. In cases where awareness is detected at a very early stage, we need to ask whether an anesthetist would be justified in infusing an amnesic drug to prevent consolidation of a memory of being aware without the patient's prior consent.

I consider whether an anesthetist would be justified or obligated to preoperatively inform patients of the possibility of awareness and such a preventive intervention as part of the consent process. When a patient forms a memory and recalls being aware, there may be reasons for and against taking a drug that hypothetically could erase it. These issues are complicated by studies indicating that anesthesia can have different effects on episodic and fear memories. They are also complicated by research showing that the longer a memory has been consolidated and stored in the brain, the more difficult it is to weaken or erase it. In addition, patients may form implicit memories that are immune to the amnesic effects of anesthesia even if they are not aware during surgery. Explicit and implicit memories formed during surgery may have long-term harmful effects on patients.

In Chapter 4, I discuss disorders of memory content and interventions to treat them. These include psychiatric disorders such as PTSD, anxiety, depression, phobia and panic. They may develop from anesthesia awareness with recall or from other factors. The emotionally charged content of a fear memory of a stressful or traumatic event persists in the brain and mind beyond any adaptive purpose. Behavioral techniques such as extinction training, and pharmacological interventions such as propranolol, have been used to weaken the emotional content of these memories or replace them with other memories. But these interventions would not rule out the possibility of reactivation of the emotional representation. Protein synthesis inhibitors infused into the basolateral amygdala may block reconsolidation and effectively erase these memories. More invasive deep brain stimulation or focused ultrasound of localized nuclei in this brain region might also erase them. Even if these interventions could erase a pathological fear memory, they would have to be selective enough not to affect normal semantic, episodic and emotional memories. Because this is still very much hypothetical, it is not known how selective memory erasure could be and whether both maladaptive and adaptive memories would be affected.

I explain how erasing some memories would not necessarily alter identity and spell out differences in the ethical justification of erasing traumatic versus unpleasant memories. Memory modification can be consistent with authenticity in a person who decides to undergo it. Just as the content of our memories is constantly being updated, so too our authentic selves are never complete but constantly being revised and reshaped by us in our interaction with the environment. Some types of memory modification may be compatible with and complement natural memory updating as part of its adaptive purpose. However, it is not known what the short- and long-term neurophysiological and psychological effects of tinkering with memory would be. This underscores the need for placebo-controlled studies to determine the feasibility, safety and efficacy of erasing memories and the circumstances in which it would or would not be justified.

I discuss disorders of memory capacity and interventions designed to treat them in Chapter 5 (Kopelman, 2002; Fradera and Kopelman, 2009). These include chronic disorders such as anterograde and retrograde amnesia and progressive aphasia, as well as temporary disorders such as transient global amnesia. Disorders of memory deficit and memory excess fall within this category. The exceptional autobiographical and semantic memory in hyperthymesia can be more of a burden than a blessing. Indeed, "hyper" suggests that the condition is abnormal and possibly pathological. Cases of a deficit or excess of memory show that optimal levels of memory are necessary for flexible and adaptive behavior. Memory stores well above or below these levels can impair or undermine this behavior. These disorders most often involve episodic memory, but also semantic, spatial, working and prospective memory. They may involve dysfunction in encoding, consolidation, storage, retrieval or reconsolidation. I consider psychotropic drugs and neurostimulation to improve, or enhance, memory. To be adaptive, enhancement should result in optimal levels of memory formation, storage and retrieval and a balance between learning and forgetting. I give examples showing how memory enhancement may involve trade-offs between learning new information and applying existing information, given that the brain can process only a certain amount of information at any given time.

In addition, I discuss how a hippocampal neural prosthetic can assist and possibly replace a damaged hippocampal-entorhinal or hippocampal-fornix circuit and improve or restore the capacity for encoding and retrieving episodic, spatial and working memory. While this brain implant has the potential to ameliorate anterograde and retrograde amnesia, there are questions about how it would interact with other subcortical and cortical circuits mediating other types of memory. It is not known whether it could replicate the function of natural hippocampal place cells and entorhinal grid cells in spatial memory. Nor is it known whether it could enable the mental time travel and autobiography that one experiences with a natural hippocampus. In addition, there are questions about the safety of a bio-mimetic device in its effects on brain tissue and the extent to which patients could control it.

In Chapter 6, I examine legal issues regarding memory. I consider how different forms of memory impairment and its effects on agency can influence judgments of criminal responsibility for actions and omissions. A person may be criminally responsible for an action he cannot recall performing. What matters in determining responsibility, mitigation or excuse is not whether episodic memory following an action is intact or fragmented. Rather, what matters is the level of working and prospective memory and how this enables or disables reasoning and decision-making at the time of action. Dissociative amnesia may impair both episodic and working memory. But there may be cases of agents in dissociative states who have enough capacity for intentional agency to be at least partly responsible for their actions. I examine cases of criminal negligence causing death and the role of memory lapses in them. The presumed cognitive control an agent has over a probable sequence of events can make responsibility transfer from an earlier action to a later consequence of the action. Assessing responsibility for omissions in criminal negligence can be difficult because of difficulty in distinguishing between a memory lapse caused by a neurological or mental disorder beyond the agent's control and a lapse caused by failure to attend to the circumstances of the action.

I also consider whether a victim of a crime would have the right to erase a memory of her experience or be obligated to retain a memory of it to testify against the perpetrator in a court of law. The fact that these memories can cause significant psychological harm and become more generalized and less accurate over time may weaken any claim of an obligation for the victim to retain the memory for this purpose. This points to a potential conflict between the disvalue of a traumatic memory to a victim of a crime and the value of the memory to society in preventing additional crimes. Describing the circumstances of a criminal act at some time later than the act may leave out critical details necessary to convict an offender. This is a practical implication of the process of updating memories. It can weaken or undermine their reliability in eyewitness testimony.

In the Epilogue, I summarize the main points in the preceding six chapters. I emphasize that adaptive memory requires a balance of remembering and forgetting and optimal levels of information in the brain and mind. This enables flexible thought and behavior necessary to meet the cognitive and emotional demands of the natural and social environment. I explore possible future directions in memory research. They include downloading information associated with memories from the brain to digital repositories external to it. They also include large-scale artificial memory systems. How these systems integrate into or replace a natural brain could influence how we assess behavior and the sorts of beings we take ourselves to be. I end by describing how information technology has expanded the concept of memory and exploring some of the social implications of this expansion. Digital memory is fundamentally different from the neurobiological and psychological processes of encoding, consolidating, storing, retrieving and reconsolidating episodic and semantic memories. The digital universe may have an unlimited capacity to acquire, store and access facts about our lives. This raises questions about whether we can draw a clear boundary between private and public information about ourselves. It expands our understanding of memory as not only an internal and individual system of information acquisition, processing and use but one that is also external and social. Unlike the episodic information in our psychological lives, the semantic information about our biographical lives may be permanently preserved and persist beyond our death.

1   Memory Systems and Memory Stages

The concept of human memory has evolved over the centuries from the simple mental act of conscious recollection to a multidimensional process involving conscious and unconscious cognitive, emotional and motor functions. Memory is not a unitary system but a collection of multiple systems. In some respects, memory systems are dissociable and can function independently of each other. Some people with anterograde and temporally graded retrograde episodic amnesia can retrieve memories of facts despite being unable to form and retrieve memories of events. This suggests that the brain systems mediating semantic and episodic memory can function independently of each other. In addition, after an initial period of conscious learning, procedural memory enabling motor skills does not rely on declarative memory. In other respects, each memory system interacts with other systems in regulating human thought and behavior. Complex action plans may require procedural, episodic, semantic, working and prospective memory. Whether memory systems are dissociable or interdependent depends on the functions at issue and the extent to which these functions and the neurobiological mechanisms that underlie them interact.

In this chapter, I trace the evolution of human memory from Aristotle to current theories of memory as a constructive and reconstructive process. The content of memory is constantly updated to enable us to meet the mental and physical demands of the natural and social environment. Although concepts of memory from Aristotle and Augustine may seem relatively simple by today's standards, they contain elements that are precursors of how researchers now conceive of memory. I mention Frederic Bartlett's research as a critical transition point in our understanding of this process. Bartlett changed the concept of episodic memory from a reproduction to a reconstruction of the past to make sense of one's actual and future experience. The emphasis on the dynamic process of updating content in current theories of reconstructive memory is a further development of Bartlett's reconceptualization of memory from his predecessors.

I describe the taxonomy of declarative (explicit) and nondeclarative (implicit) memory systems and explain the different respects in which they enable cognitive, emotional and motor functions. Although these systems are often considered independently of each other in corresponding to conscious and unconscious functions, I cite examples of how they interact. I also describe the process extending from encoding and storing short-term memory to consolidating, retrieving, reconsolidating and storing long-term memory and the neurobiological structures and mechanisms that regulate this process. I explain how different disorders of memory are caused by neural and mental dysfunction affecting one or more of these stages. These disorders typically involve the declarative memory subtypes of episodic and semantic memory. The taxonomy of memory systems and the structure of the different stages in the memory process will serve as a framework within which to analyze and discuss how memory function and dysfunction influences our understanding of agency, identity and responsibility. It will serve as a framework for exploring the metaphysical, ethical and legal implications of these concepts and interventions to improve, restore, weaken or erase memories.

History

The first systematic treatment of memory in Western thought was Aristotle's *On Memory and Recollection* (c. 350 BCE; Aristotle, 1984). Aristotle was particularly interested in recollection. This process involved a succession of associated ideas about events in the past. In the *Meno* and *Phaedo* (c. 400 BCE), Plato claimed that recollection was the route to knowledge of the Forms, the exclusive objects of knowledge (Plato, 1962). In addition, he argued that recollection was a way of proving the existence of the soul before birth. Plato also claimed that dialectical discussion required participants to recollect relevant data (Sorabji, 2004, p. 36). Aristotle rejected Plato's metaphysical views about the Forms and the soul. Yet he was interested in the connection Plato drew between recollection and dialectic. Richard Sorabji notes that Aristotle takes this connection one step further. "He thinks the ability to recollect is so important for successful dialectical debating, that he urges his students to memorize certain things, in order to prepare themselves for such debates" (p. 36). This could be described as the first discussion of semantic memory of facts and concepts about the world. It could also be described as the first discussion of working memory in executive tasks like debating, which draw on semantic stores in holding and processing information for brief periods. Aristotle's conception of recollection is different from George Mandler's division of recognition memory into the dual-process retrieval model of faster, less cognitively demanding, familiarity and slower, more cognitively demanding, recollection (Mandler, 1980).

In his *Confessions* (397–400 CE), Augustine made two contributions to the conceptualization of memory that marked a major advance over Aristotle's account of recollection. In this work, Augustine distinguished between the act of remembering and what we remember (Augustine, 2008, Bk. X, ch. xiv, Bk. XI, ch. xviii). This involves what we would describe as the distinction between memory retrieval and the content or representation of what is retrieved. His second and equally significant contribution was his emphasis on the connection between time and memory. This may have been the first discussion of "mental time travel," later associated with Endel Tulving (Tulving, 1983, 2005). It is a precursor of what Tulving called "autonoetic consciousness," or diachronic awareness of the self in subjective time. This conscious capacity to mentally place ourselves in the past, present and future is essential to autobiographical memory as a subtype of episodic memory (Tulving, 1983, 1985b, 2005; Manning, Cassel and Cassel, 2013). Augustine writes, "The time present of things past is memory; the time present of things present is direct experience; the time present of things future is expectation" (2008, Bk. XI, ch. xx). Although Augustine describes memory as a backward-looking capacity in this passage, the connection he draws between memory, experience and expectation establishes a diachronic concept of consciousness based on memory.

Thirteen centuries after Augustine, in *An Essay Concerning Human Understanding*, John Locke developed an account of memory as a form of mental storage. Memory was a critical component of Locke's account of personal identity in linking one's present to one's past. "This is memory, which is as it were the store-house of ideas. For the narrow mind of man, not being capable of having many Ideas under View and Consideration at once, it was necessary to have a Repository, to lay up those Ideas, which at another time it might have use of" (Locke, 1690/ 1975, Bk. II, ch. x). The "store-house" for Locke is not a static repository but a dynamic one. Ideas, including memories, exist only when they are perceived consciously by the mind. Although Locke did not explain it in these terms, his use of "perceive" and "store" may be a precursor of the current understanding of retrieval, reconsolidation and long-term storage of memory. Without being perceived (retrieved), episodic and semantic memories would not be reconsolidated and stored in the brain and mind and eventually would be removed from them.

In the late nineteenth century, Hermann Ebbinghaus retained the basic concept of memory as a storehouse of ideas and a reproduction of experience (Ebbinghaus, 1885/1913). Equally significant was the distinction

Ebbinghaus drew between short-term and long-term memory, and how the second form depends on repeated retrievals of the mental representation associated with the memory (Squire and Kandel, 2009, p. 4). This distinction was similar to William James's distinction between primary and secondary memory in his *Principles of Psychology* (James, 1890; Eichenbaum, 2012, p. 15). The inclusion of both the content of what we learn and the experience of learning in secondary, long-term, memory was a further development of Augustine's distinction between the act of remembering and what we remember.

A fundamental change in the conceptualization of memory occurred in 1932 with the publication of Frederic Bartlett's *Remembering*. According to Bartlett's "schema theory," remembering is a reconstructive rather than reproductive process, as Ebbinghaus conceived of it. Declarative memory is "an imaginative reconstruction, or construction, built out of the relation to our attitudes towards a whole active mass of organized past reactions or experience, and to a little outstanding detail which commonly appears in image or in language form" (1932, p. 213). Memory for Bartlett is more than a cognitive function of bringing information about the past to the mind's eye. It is imbued with emotion and meaning as a reflection of the natural and social context in which the person lives and acts. Memory shapes and is shaped by the integration of past, present and future dimensions of a person's life. Bartlett rejected the idea of memory as a reproduction of past events occurring in a vacuum. He believed that "it was impossible to separate the acquisition and recovery of memory of an experienced event from the knowledge, motivation and culture of the individual. Memory was deeply embedded in the individual's pre-existing knowledge structures, or 'schemas,' and inseparable from them" (Eichenbaum, 2012, pp. 82–83; Moscovitch, 2012, p. 22). Bartlett's reconstructive model is the basis of what has become the consensus view in contemporary memory theory.

Defining Memory

Memory is a reexperiencing of events. As Tulving describes it, episodic memory involves the capacity to project oneself backward to experienced events (1983, 2000). It involves imprinting, retaining and applying information about the past to execute tasks requiring cognitive, emotional and motor capacities in the present and future. Morris Moscovitch writes, "Memory is a lasting, internal representation of a past event or experience (or some aspect of it) *that is reflected in thought or behavior*. It follows, therefore, that *memory does not exist until it is recovered*" (2007, p. 17). This is similar to Locke's account of memories and other ideas that exist only when a person consciously perceives them during retrieval. Augustine's emphasis on the act of remembering is pertinent to this definition as well. It points to the critical role of retrieval of episodic and semantic memory in allowing for reconsolidation and retaining it as information in the brain and mind.

The degree to which we integrate and store information about the past is also a product of genetic, epigenetic and neurophysiological mechanisms (Kandel, 2001). These include the transcription factor cyclic-AMP response element binding protein (CREB), protein synthesis and long-term potentiation (LTP). This is a form of synaptic strengthening that refers to long-lasting facilitation of synaptic transmission necessary for memory formation. How memories are consolidated and reconsolidated depends on how LTP and the action of N-methyl-D-aspartate (NMDA) receptors affect the rate of neural firing and synaptic connectivity (Kandel, 2001; Li and Tsien, 2009; Squire and Kandel, 2009, ch. 6). There are differences in how these mechanisms enable different types and stages of memory. Some may be more important than others in consolidation and reconsolidation, or storage and retrieval. But they all have a critical role in all of these stages. The stronger the degree of neural firing and synaptic connectivity in hippocampal-cortical pathways, the more likely memory will remain stored and available for retrieval. The weaker the degree of firing and connectivity, the less likely the memory will remain stored and retrieved. Beyond a certain point, weakened neural firing and synaptic connectivity will result in reduced integration or loss of information associated with the memory.

The terms "engram," "trace" and "representation" are often used interchangeably in accounts of memory. Engram is defined as the representation of the memory in the brain rather than the memory itself. Moscovitch explains that "the engram or memory trace is the representation of an encoded event or experience. It is not yet a memory but provides the necessary (physical) condition for memories to emerge. Put another way, the engram is permissive or necessary for memory but does not suffice for its materialization. A memory emerges when the engram interacts with retrieval cues or information derived from particular environmental conditions" (2007, p. 18). Tulving defines "trace" as the neural change that occurs in the interval between a mental experience and recalling that experience (2007, p. 67). This change involves the process of encoding and consolidating information. According to the Multiple Trace Theory (MTT) of memory consolidation, information about an event or fact is encoded and consolidated in a unique memory trace (Nadel and Moscovitch, 1997; Moscovitch and Nadel, 1998; Nadel, Hupbach, Gomez et al., 2012).

Moscovitch's and Tulving's definitions indicate that memories are not concrete objects contained in or identical to brain tissue. They are emergent properties associated with information in the brain and how synaptic connectivity integrates this information. Memories emerge from this information when it reaches a critical level of integration. Memories are not "snapshots" of events. They are neurally encoded representations of information associated with events. Neither in vivo nor ex vivo examination of human brain tissue could reveal a person's memories. Tulving claims that "A memory trace is a change, and a change is not an entity" (2007, p. 67). Defining "a memory trace in terms of physical changes in the brain does not make the trace a physical entity itself" (p. 67). Tulving further claims that "because there is no object or entity in the brain that can be said to *be* a particular memory trace (i.e. to represent a particular experience), it is in principle not possible to observe it as such, to identify it as such or to determine its properties as such" (p. 67). The memory trace is not identical to one or more synapses or neurons in a specific brain region. Rather, their excitatory activity and firing patterns make the trace possible. This prompts the question of how we can know that a person forms, stores, retrieves or loses memories if they are not identical to any structural or functional features of the brain.

These processes cannot be known directly by observing a localized area of neural tissue. They can only be known indirectly. They can be inferred from changes in synaptic connectivity indicating a change from a more inhibitory to a more excitatory state, or vice versa. This activity can be detected by functional imaging such as positron emission tomography (PET) or functional magnetic resonance imaging (fMRI), together with changes in the person's thought and behavior. It corresponds to changes in information processing from a less integrated to a more integrated state, or vice versa. This provides a neuroscientifically informed answer to the questions of "where" memories are formed and stored, and where they "go" when we cannot retrieve or permanently lose them in neurodegenerative disorders such as dementia. The extent of integration of neural information and the strength of synaptic connectivity are necessary to explain how a memory can be weakened or erased by extinction training, a beta-adrenergic receptor antagonist, a protein synthesis inhibitor, high-frequency electrical brain stimulation or high-intensity focused ultrasound.

Following his earlier three-fold distinction between episodic, semantic and procedural memory (Tulving, 1983, 1985a, 1985b, 1987), Tulving later distinguished six different sense of "memory" (2000, pp. 33–44). Daniel Schacter cites four of these senses as particularly relevant to thought and behavior: (1) the neurocognitive capacity to encode, store and retrieve information, (2) a hypothetical store in which information is held, (3) the information in that store and (4) an individual's phenomenal awareness of remembering something (Schacter, 2007, p. 25). These interrelated senses of memory, and particularly item 4, support the claim that memory is not just a neurobiological or psychological process but is both neurobiological and psychological. The phenomenal awareness of memory involves a subjective aspect that cannot be explained entirely in terms of neural firing and synaptic connectivity. Nor can brain mechanisms alone explain the content of memory, the events, facts and experiences of which we have representations. The phenomenological aspect of episodic memory corresponds to Tulving's characterization of the autonoetic aspect of this memory type.

These four senses of memory suggest a model in which brain and mind are not distinct substances but interacting and interdependent processes. They suggest that the theory that most accurately captures how brain and mind interact in generating and sustaining memory is nether substance dualism nor reductive materialism but nonreductive materialism (Baker, 2013, part II; Northoff, 2014, pp. 100ff.). A normally functioning hippocampal complex and its connections to frontal, temporal and parietal cortices is necessary to encode, store and retrieve declarative memories. These capacities may be impaired by brain injury or neurological disease. They may also be impaired by psychological processes. Acute emotional distress or psychosocial stress may cause transient global amnesia by impairing consolidation and retrieval mechanisms in the hippocampus and neocortex (Zeman, Kapur and Jones-Gotman, 2012a, p. 9). Consistent with nonreductive materialism, psychogenic memory disorders indicate that mental states can causally influence neural states. The experience of having a memory is related to a neural change and cannot exist in the absence of this change (Tulving, 2007, p. 68). But the subjective quality and content of the memory are not reducible to neural processes. They are also shaped by contextual factors outside the brain and the person's own unique experiential history. This is consistent with Bartlett's view that it is impossible to separate memory from the "schemas," the knowledge, motivation and culture of the individual (Moscovitch, 2012, p. 22)

The Constructive and Reconstructive Model of Memory

Many psychologists and some philosophers have endorsed a constructive, or reconstructive, model of episodic memory. In the passage cited from *Remembering*, Bartlett uses "reconstruction" and "construction" interchangeably. These are closely related processes. Schacter and Donna Rose Addis state that episodic memory "is not a literal reproduction of the past but is instead constructed by pulling together pieces of information from different sources" (2007a, p. 27). We do not literally create this information. The brain encodes and consolidates the information that becomes the original trace. We then update the content of the trace through the combined action of our minds and brains in retrieving and reconsolidating the memory. Updating makes memories meaningful and adaptive. We reconstruct their content when we update them. These points suggest that it may be appropriate to describe memory as both a constructive and reconstructive process.

Chris Westbury and Daniel Dennett claim that "what we recall is not what we actually experienced but rather a reconstruction of what we experienced which is consistent with our current goals and our knowledge of the world" (2000, p. 19). Similarly, Schacter and Addis point out that the purpose of memory is not to form an accurate representation of the past but to enable us to mentally simulate possible episodes in our personal experience (Schacter and Addis, 2007a, 2007b; Schacter, Addis and Buckner, 2008). They claim that "information about the past is useful only to the extent that it allows us to anticipate what may happen in the future" (Schacter and Addis, 2007b, p. 27; Schacter, 2012). We use information about the past to construct a model for goal-directed behavior. This forward-looking concept of episodic memory complements Tulving's backward-looking concept. Combined, these two perspectives provide a diachronic sense of experience. The future is conceived of as a mirror image of the past in the sense that we use it to imagine possible states of affairs and courses of action (Addis and Schacter, 2012; Schacter, Addis, Hassabis et al., 2012; De Brigard, 2013, 2017; De Brigard, Addis, Ford et al., 2013). By informing action plans, memory enables effective agency. As consciously accessible information about experienced events, "episodic memory allows us to employ representations of these events in the service of current and future goals" (Rugg and Vilberg, 2013, p. 255). Because information about the past influences how we form and revise these goals, memory enables flexible, adaptive behavior (Suddendorf and Corballis, 2007).

Too much information about the past can clutter the brain and mind and interfere with the ability to execute immediate and future cognitive tasks. Effective agency requires a balance between remembering and forgetting. Schacter and Addis further note, "Many researchers believe that remembering the gist of what happened is an economical way of storing the most important aspects of our experience without cluttering memory with trivial details" (2007b, p. 27). Agreeing with this position, they also claim that, according to the constructive model, "memory is important for the future as well as the past" (p. 27). "Gist" suggests that at least some of the original content of the memory is preserved in long-term memory storage for use in projecting ourselves into and planning for the future. "Trivial details" may be associated with episodic memory of events or semantic memory of facts. To avoid overloading the brain and mind with irrelevant information, repeated retrieval and reconsolidation of episodic memories causes them to become less detailed and more generalized. They become "gist-like semanticized memories" (Moscovitch, 2012, p. 31). Semantic as well as episodic information has a role in constructive memory (Irish and Piguet, 2013). This process has important implications for the role of memory in agency. Retrieval and reconsolidation involve updating the content of episodic and semantic memory to reflect a person's circumstances. These processes enable the person to alter action plans to align with these circumstances (Nader, Schafe and LeDoux, 2000; Nader and Einarsson, 2010; Nader, Hardt, Einarsson et al., 2013). Updating memories enables us to adapt to changing natural and social environments.

The idea that memory does not require an exact representation of experience to promote goal-directed behavior suggests that some degree of misremembering may occur in the constructive process (Schacter, Guerin and St. Jacques, 2011; Schacter and Loftus, 2013). Indeed, updating memories may require some misremembering. Nevertheless, at least some of the original content of the memory must be preserved as an informational basis to guide current and future behavior. There cannot be so much misremembering that one loses the gist of what happened. There are limits to how much the original content of the memory can be altered for it to be adaptive.

Reconstructive memory is analogous in some respects to the antifoundationalism in epistemology expressed by Otto Neurath in his ship metaphor: "We are like sailors who, on the open sea, must reconstruct their ship but are never able to start afresh from the bottom … by using the old beams and driftwood, the ship can be shaped entirely anew but only by gradual reconstruction" (Neurath, 1921, p. 191; Cartwright, Cat, Fleck et al., 1996, pp. 89ff.). Unlike Neurath's ship, reconstructed memory is not "shaped entirely anew" but retains some of the content of the original trace. While details of the past can interfere with the capacity to simulate alternative possibilities of action and project ourselves into the future, the core features of our experience and what we learn from it shape these capacities. A scaffolding of a less detailed and more generalized set of memories from past episodes is necessary to reconstruct them so that they are relevant to our present and future circumstances. More distant episodic memories in this scaffolding are gradually replaced by more recent memories. But there must always be a core set of memories with a certain degree of stable content to plan, act and integrate our life experiences into a meaningful whole. The longer the period between an original experience and later retrieval of a memory of that experience, the weaker the memory trace becomes until it fades and is removed from storage. When the brain functions normally, the length of time that memories are stored and available for retrieval depends on their relevance to the cognitive, emotional and motor capacities necessary to successfully navigate the environment.

These considerations suggest that the most plausible account of memory may be one consisting of a hybrid of reproduction and reconstruction models. Moscovitch endorses this position: "Given our current state of knowledge, perhaps it is best to adopt a middle ground between Ebbinghaus and Bartlett. Recovering the past and imagining the future involves the interaction of processes or memory reproduction and reconstruction, which are dependent on a core of relational memory traces mediated by the hippocampus (Addis and Schacter, 2012). Indeed, recent work on reconsolidation in animals … suggests that memories may never fully be consolidated but need constantly to be reactivated to be maintained … during reactivation they are susceptible to modification, interference and loss, processes that can occur throughout one's lifetime" (Moscovitch, 2012, p. 39). This underscores the idea of memory as a dynamic process that is inseparable from one's diachronic lived experience. Again, though, adaptability requires retention of the core content – the "gist" – of a set of original experiences and long-term memories of them. A storehouse that is thinner than what Locke and Ebbinghaus envisaged may be sufficient to preserve this content. Only a certain amount of information about the past needs to be preserved and reproduced as a basis for memory construction and reconstruction. But a foundational store is necessary to respond to present demands and anticipate the future.

The constructive model of memory pertains only to episodic and semantic subtypes of declarative memory. It does not provide a comprehensive model of all memory systems. Although working memory is a necessary basis for future planning, we do not construct this memory when we use it. Instead, we use information temporarily available to us to execute immediate cognitive tasks. Nor do we consciously alter the content of our experience when we unconsciously use procedural memory to automatically perform motor tasks. Nevertheless, constructivism and reconstructivism are significant because of the extent to which episodic and semantic memory regulate our lives. Updating episodic memory is critical to agency because the capacity to revise information about the past to make it relevant to present and future circumstances is necessary to form and complete action plans.

This process is also relevant to personal identity in allowing us to develop a unified narrative of our experience of persisting through time. Our selves are not fixed and complete. We constantly shape and reshape them as we update the content of our episodic memories in response to the natural and social environment. This allows the content to align with our actual and possible experience. The process of replacing older memories with more recent memories in the updating process underscores the importance of optimal levels of information in the brain. Too much information, as in hyperthymesia (from the Greek: *thymesis* = memory), or too little information, as in amnesia, can result in maladaptive behavior. Constructivism is also relevant to semantic memory and how we use our knowledge of facts and concepts to model and interact with the external world.

The role of sleep in memory further illustrates its constructive aspect. Sleep is critical to the adaptive function of memory and its meaning for the organism and subject. Consistent with the idea of balance between learning and forgetting, a balance between high-amplitude, low-frequency noise in NREM sleep and low-amplitude, high-frequency noise in REM sleep leads to the elimination of all but the strongest memory traces. Structural changes in the prefrontal region of the aging brain can disrupt this balance and interfere with the brain's ability to consolidate and store new memories (Mander, Rao, Lu et al., 2013). NREM–REM balance during sleep is one way of removing memories that are irrelevant to or interfere with one's current cognitive demands. It prevents information overload in the brain and mind. Normal cycles of NREM and REM sleep thus promote flexible thought and behavior.

Robert Stickgold explains that "the construction of 'meaning' refers to the integration of new memories into pre-existing memory networks to provide an enhanced context from within which pre-existing information can better inform future action. From a phenomenological perspective, this same integration provides a context that facilitates recognition of associations and relationships between new memories and elements in these networks, and that promotes identification of regularities of which the new memory is a part, and of rules that apply to it" (2011, p. 77). Much of this integrative process occurs within cortical networks in which memories are stored following consolidation and re-stored following reconsolidation. The organization of information in these networks makes it meaningful to the person by enabling her to understand her relation to the world. This process is not limited to episodic memories. Sleep also enhances consolidation of motor, verbal and emotional memories (Maquet, 2000; Stickgold, 2011, p. 82). Particularly significant is how "sleep contributes to the slow evolution of memories, from clear and accurate records of recent events into generalized memories of the gist of events, placed within the context of memories from disparate times, and used to develop new conceptual memories describing the rules and meanings of events in our lives" (Stickgold, 2011, p. 82).

Sleep has an essential role in the transition from detailed episodic memories of experienced events to general gist-like semanticized memories of these events. This contextualization of memories provides us with the basic information about the past that we need for planning and acting. It also spares the brain and mind from having to process large volumes of information and avoids cognitive overload that would interfere with our ability to make sense of this information. "Sleep appears to be a state in which the brain specifically acts to construct the meaning of events in our lives" (Stickgold, 2011, p. 85). Given the function of the hypothalamus in regulating sleep, the connection between sleep and memory indicates a connection between the hippocampal complex and the hypothalamus in the memory process. The role of sleep in memory is another example of how neural networks regulate the conversion of information into meaningful memories.

Sleep disorders may be linked to semantic amnesia, the inability to recall and use words, numbers and concepts. Gabriel Garcia Marquez uses fiction to illustrate this link in his novel, *One Hundred Years of Solitude* (Garcia Marquez, 1970). Following a plague of insomnia, the inhabitants of Macondo have difficulty recalling words and fall into a collective semantic amnesia. "One day, the village blacksmith was looking for the small anvil that he used for laminating metals, and he could not remember its name … This was the first manifestation of a loss of memory, because the object had a difficult name to remember." Eventually, the people sink "irrevocably into the quicksand of forgetfulness" (Garcia Marquez, 1970, p. 54). Fortunately, a cure is found for the illness. Katya Rascovsky and coauthors explain that the villagers' symptoms correspond to the neurological disorder of semantic dementia (SD). They note that "the cognitive impairments experienced by Macondo's inhabitants are remarkably similar to those observed in SD, a clinical syndrome characterized by a progressive breakdown of conceptual knowledge (semantic memory) in the context of relatively preserved day-to-day (episodic) memory. First recognized in 1975, SD is now considered one of the main variants of frontotemporal lobar degeneration" (Rascovsky, Growden, Pardo et al., 2009, p. 2609). Garcia Marquez suggests that semantic amnesia could be the result of severe insomnia as well as a neurodegenerative disorder. Regardless of the cause, people who suffer from semantic amnesia are not as fortunate as the inhabitants of the fictional Macondo, given the lack of safe and effective treatment for this memory disorder (Matthews, 2011, p. 118).

How many episodic and semantic memories need to be stored in the brain? *How long* do they need to be stored to enable goal-directed behavior? Because we cannot predict the future, it is not known how much or what type of information about the past we need to adapt to events that we might or might not experience. A sufficient store of older memories should be available to us to respond cognitively and emotionally to natural and social contingencies. But there cannot be so many older memories that they limit the consolidation and storage of newer memories that may be just as relevant to our need and ability to meet these demands. Research suggests that the brain has mechanisms for selecting and retaining information that is more likely to promote flexible behavior in future environments (Dunsmoor, Murty, Davachi et al., 2015). The brain suppresses retrieval of more distant memories so that they do not interfere with the role of more recent memories in effective planning for the future (Anderson, 2014; Wimber, Alink, Charest et al., 2015). This is an example of how a certain degree of forgetting can be adaptive.

How can we distinguish misremembering as an adaptive function from misremembering as a maladaptive symptom of brain dysfunction from brain injury or dementia? An answer to this question depends on more than what neuroimaging can detect about brain structure and function. Intuitively, if misremembering does not interfere but is compatible with or promotes the person's ability to plan for the immediate or near future, then it is adaptive. If it interferes with this capacity, then it is maladaptive. In cases of people with severe retrograde amnesia, brain damage disrupting functional connections between the hippocampus and regions of the frontal, temporal and parietal cortices can impair or undermine the capacity to imagine the future. This capacity relies on retrieval of long-term memory, which is impaired or lost in this memory disorder. It underscores the point about the need for a constant store of the core content of episodic and semantic memory through changes from repeated retrieval and reconsolidation. Although misremembering may be caused by brain dysfunction, whether it is adaptive or maladaptive ultimately depends on the person's behavior. This does not answer all questions about the accuracy of memory, however.

True, False and Imaginary Memories

One of the most challenging issues in memory research is distinguishing between true, false and imaginary memories. True memories accurately represent a person's actual experience of actual events. False memories have no connection to this experience or these events. More difficult to distinguish are true and imaginary memories because the subjective aspect of reconstructing the past may either update or completely change its content. Daniel Bernstein and Elizabeth Loftus claim that "All memory is false to some degree. Memory is inherently a reconstructive process, whereby we piece together the past to form a coherent narrative that becomes our autobiography. In the process of reconstructing the past, we color and shape our life's experiences based on what we know about the world" (2009, p. 373). This suggests that memories are neither strictly true nor false but instead have varying degrees of reliability and utility. For those who endorse constructivism in memory, the key issue is not so much whether memories are true or false but more so their instrumental value in promoting adaptability and allowing us to derive meaning from our experience. Still, some will insist that memories must be veridical to be adaptive. Memories must be connected to actual events to provide a stable basis for the mental act of simulating possible states of affairs and courses of action.

These considerations point to the need for a causal theory of constructive memory (Michaelian, 2016, pp. 80ff.). This "causal theory distinguishes remembering from imagining by requiring a continuous causal connection running from the subject's original experience of the relevant event to his retrieved representation of the event" (p. 80). More precisely, the causal connection holds from the experience of the event, through the encoding, consolidation and storage of its representation, to later retrieval, reconsolidation and re-storage of the representation. The sequence of events presupposes normal neural and mental function. False or imaginary memories could not be part of the causal chain if the origin of the experience and trace were a brain abnormality or an artificially induced neural and mental state. The sequence of events could be disrupted by abnormalities in one or more stages of the memory process. The causal intermediaries between an experience and subsequent activation of the memory trace must avoid these conditions to be connected in the right way. This allows the content of the memory at the earlier time to be suitably related to its content at the later time. Equally significant, the source of the memory must be an actual event or experience that the subject can report to others and is objectively verifiable.

Roughly, true memories accurately represent a person's experience of actual events in the external world. False memories have no connection to this experience or these events. Whereas the content of true memories is information about the past, the content of imaginary memories is the simulation of future possibilities. But remembering the past and imagining the future are two sides of the same coin. We use our knowledge of the past to project ourselves into the future. Without a basis in actual facts or events, imaginary memories can become false memories.

Neuroimaging studies have helped to distinguish between recalling actual past events and imagining possible future events by showing different patterns of neural activation. They have shown that "the right hippocampus was engaged to a greater extent when participants imagined future events than when they remembered past events" (Squire, van der Horst, McDuff et al., 2010; Schacter, Chamberlain, Gaesser et al., 2012, p. 246). These patterns of brain activity reflect the additional relational processing involved in imagining and "the more intensive constructive processes required by imagining future events relative to retrieving past events" (Schacter, Chamberlain, Gaesser et al., 2012, p. 246). The results of these studies "appear to provide evidence that right hippocampal activation constitutes a neural signature associated with the construction and encoding of specific imagined events" (p. 247). Imaging-based differences in hippocampal activity when subjects are asked to recall actual events or imagine possible events provide empirical support for the claim that true and imaginary memories can be described as complementary functions of past- and future-oriented neural and mental processes.

But neuroimaging alone cannot determine that a memory is true or false because investigators can manipulate brain function to artificially induce memories. One group of researchers implanted false memories in the brains of sleeping mice using optogenetics in manipulating light-sensitive proteins (De Lavilleon, Lacroix, Rondi-Reig et al., 2015). It is possible that this and other techniques could implant false memories in humans as well. These memories are false in the sense that they have no causal connection to stimuli or events external to the brain. They are entirely the product of neural manipulation. Yet the information associated with these memories could elicit apparently normal patterns of hippocampal activation.

Out-of-body experiences (OBEs) or near-death experiences (NDEs) in humans present a different challenge for the reliability of memories because people report having them. Presumably, the reports imply having a memory of the experience. Some researchers have attributed OBEs to brain dysfunction in the temporal-parietal junction (Blanke, Landis, Spinelli et al., 2004). These experiences can also be induced by electrically stimulating this brain region. Other researchers have found that high-frequency neurophysiological activity during cardiac arrest can exceed the normal level during the conscious waking state and generate NDEs (Borjigin, Lee, Liu et al., 2013). In a different study of 140 survivors of cardiac arrest, 2 percent of those interviewed reported awareness with explicit recall of "seeing" and "hearing" actual events related to the resuscitation (Parnia, Spearpoint, de Vos et al., 2014). Just because the experience and recall of it was associated with heightened neuro-physiological activity does not mean that the memory was false. If the memory was of events related to the resuscitation external to the patient's brain, and if these events could be validated by the medical team performing the resuscitation, then the memory would be true. It would be a representation of a real event rather than an imagined one. Its content would be based on verifiable events. Although neural and mental reconstruction can alter the content of a memory, a verifiable causal connection between the memory and the event or events that triggered it may be enough to establish the veracity of the memory.

I base my discussion of the metaphysical, ethical and legal aspects of understanding and modifying memory on the assumption that the relevant episodic and semantic memories are representations of actual events and facts. These events and facts have representations in normally functioning neural networks and the conscious mental states in which they are recalled. While the content of memory can change following retrieval and reconsolidation, I assume that the events and facts we remember have occurred and obtain independently of the mental act of remembering and that our memories are accurate representations of them. Following Augustine, we should not conflate the act of remembering with what we remember. The content of a memory is shaped by how we initially perceive it. The reconstructive process may alter this content from its initial representation. Still, if the experience a memory represents is objectively verifiable, then it is neither false nor completely imagined.

How much of the content of an episodic memory is altered over time is not just a function of the retrieval process. It is also a function of the updating that occurs with reconsolidation following retrieval. Even with the changes associated with updating, there can be a causal connection between the content of a memory after reconsolidation and the original experience and memory of it. People with hyperactive retrieval mechanisms revise the content of their memories and are just as susceptible to imagining or reporting false memories as those with a normal rate of retrieval (Patihis, Frenda, LePort et al., 2013; LePort, Stark, McGaugh et al., 2015). It is important to emphasize, though, that whether a memory is true, false or imaginary depends not just on retrieval but also on encoding and consolidation of the information associated with it.

Questions about the veracity or reliability of memory may persist because retrieving episodic memory is not just a neurobiological process but also a psychological process. Even when a person's recall of an experience is traceable to an actual event, the act of recalling the experience has a subjective aspect that cannot be described entirely in neuro-physiological terms. This self-conscious autonoetic aspect of episodic memory is one of the features that distinguish it from semantic memory (Tulving, 1983, 2005). Both real and imagined episodic memories have a subjective aspect. Nothing about autonoesis itself points to an objective criterion to distinguish them. Moreover, there is some degree of imagination in reconstructing or revising the content of a memory. This may make it difficult to clearly separate a reconstructed memory of an actual event from an event that is merely imagined. The origin of a memory is not the only factor in its adaptive purpose. But it must be grounded in an actual past event external to the person's brain and mind to be adaptive. If a memory is traceable to a verifiable experience or event, if it is mediated by normal brain function and if there is enough of a resemblance between the original experience and the reconstructed memory of it, then there may be enough of a causal connection between them. This may provide sufficient grounding for the memory to be a reliable guide for our current and future action.

Verifiability may at times be elusive. As I discuss in Chapter 3, in some cases there may be no way of knowing whether a patient becomes aware during surgery and forms a memory of the experience. Like consciousness, memory falls along a representational continuum. Just as it may be difficult to determine whether one has a sufficient level of integrated information to be conscious, so too it may be difficult to determine whether one has a sufficient level of integrated information about an event or fact to form a veridical representation of it. In these cases, there may be some degree of uncertainty about whether a memory is true, imaginary or false.

Systems and Stages

Episodic and semantic memory are declarative, or explicit, types of memory. This is memory that is accessible to consciousness (Tulving, 1983, 1985a, 1985b, 1987; Graf and Schacter, 1985). Declarative memory can have both cognitive and emotional content. Procedural memory is one type of nondeclarative, or implicit, memory. This is memory that operates outside conscious awareness. It involves some unconscious cognitive functions but is mainly associated with motor skills. Described as different types of knowledge, episodic memory is "knowing *when*," semantic memory is "knowing *that*," and procedural memory is "knowing *how*." Larry Squire and Stuart Zola outlined a comprehensive taxonomy dividing declarative and nondeclarative memory into systems and subsystems. This taxonomy captures the multiple dimensions of human memory and its role in enabling cognitive, emotional and physical capacities (Squire and Zola, 1996, p. 13516) (see Figure 1.1). It also cites the brain regions that mediate different types of

LONG-TERM MEMORY

NONDECLARATIVE (IMPLICIT)

DECLARATIVE (EXPLICIT)   PROCEDURAL   PRIMING   SIMPLE   NONASSOCIATIVE
                         (SKILLS   CLASSICAL   LEARNING
                         AND   CONDITIONING
                         HABITS)

FACTS ——— EVENTS   EMOTIONAL   SKELETAL
                                                 RESPONSES   MUSCULATURE

MEDIAL TEMPORAL LOBE   STRIATUM   NEOCORTEX   AMYGDALA   CEREBELLUM   REFLEX
DIENCEPHALON   PATHWAYS

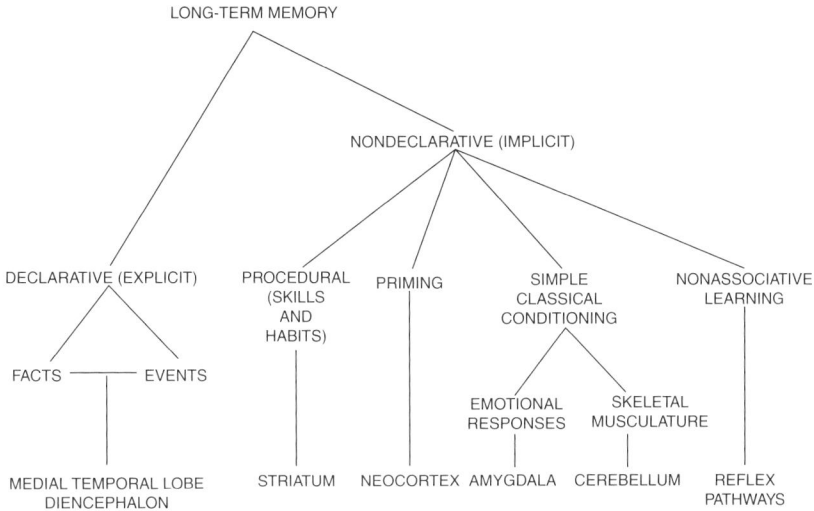

Figure 1.1 Taxonomy of the different types of declarative and nondeclarative memory and the brain regions that mediate them. From L. Squire and S. Zola (1996), Structure and function of declarative and nondeclarative memory systems, *Proceedings of the National Academy of Sciences* 93: 13515–13522, at 13516. (with permission from Larry Squire)

memory (see Figure 1.2). These systems are separate in mediating some memory functions and highly interactive and interdependent in mediating others (Eichenbaum, 2012, pp. 219ff.; Veselis, 2017).

Episodic and semantic memories are encoded and consolidated in the hippocampal complex, which consists of the hippocampal formation (hippocampus, dentate gyrus, subiculum), and the parahippocampal gyrus (entorhinal, perirhinal and parahippocampal cortices) in the medial temporal lobe (MTL). With some qualifications that I will note shortly, these memories are retrieved from short-term storage sites in the hippocampal complex and long-term storage sites in the neocortex (frontal, temporal and parietal cortices). The regions in which memories are stored correspond to the regions in which the information was first learned. The encoding and consolidation of emotional memories involves interaction between the hippocampus and amygdala (McGaugh, 2000; Richardson, Strange and Dolan, 2004). Short-term memory refers to the initial encoding and storage of declarative memories for a very brief period of approximately 30 seconds. They are then consolidated and converted to long-term memories, which is necessary for them to be retrieved from and reconsolidated in long-term storage.

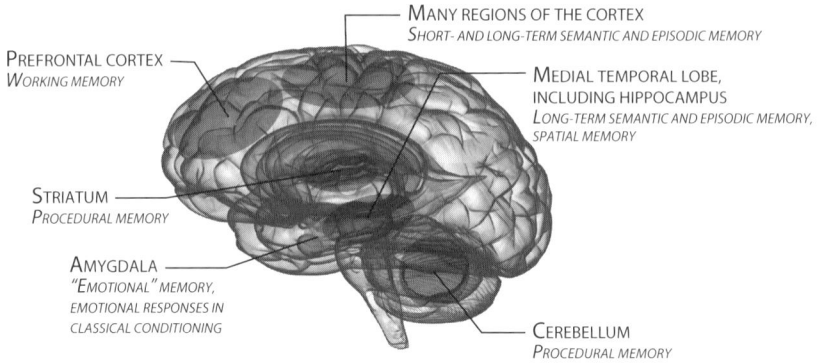

MANY REGIONS OF THE CORTEX
SHORT- AND LONG-TERM SEMANTIC AND EPISODIC MEMORY

PREFRONTAL CORTEX
WORKING MEMORY

MEDIAL TEMPORAL LOBE,
INCLUDING HIPPOCAMPUS
*LONG-TERM SEMANTIC AND EPISODIC MEMORY,
SPATIAL MEMORY*

STRIATUM
PROCEDURAL MEMORY

AMYGDALA
*"EMOTIONAL" MEMORY,
EMOTIONAL RESPONSES IN
CLASSICAL CONDITIONING*

CEREBELLUM
PROCEDURAL MEMORY

Figure 1.2 Mapping memory. The main regions of the brain mediating declarative and nondeclarative memory. Many of these regions are subcortical. Transcranial magnetic stimulation (TMS) and transcranial direct current stimulation (tDCS) only penetrate the cortex and may be limited to improving working and semantic memory. Deep brain stimulation can penetrate subcortical regions and have a broader range of neuromodulating effects on memory functions. (with permission from Christian Ineichen, image creator)

Working memory consists in processing information over brief periods (from several seconds to minutes) to perform cognitive tasks such as calculating and problem-solving. Short-term and working memory are limited-capacity types of memory. What distinguishes them is that the first type has a temporary *storage* function and the second a temporary *processing* and *executive* function (Gathercole, 2007, p. 155). Working memory draws from both short- and long-term episodic and semantic memory stores. It receives inputs from and sends inputs to long-term memory (Gathercole, 2007, p. 157). Alan Baddeley defines working memory as "a limited capacity system, which temporarily maintains and stores information and supports human thought processes by providing an interface between perception, long-term memory and action" (2003, p. 829). Information storage in working memory depends "on a memory trace that would fade within seconds unless refreshed by rehearsal" (Baddeley, 2007, p. 7). Whether this information is "refreshed" (retrieved) depends on one's momentary cognitive demands.

The hippocampal complex regulates short-term memory. Sensory, prefrontal, temporal and parietal cortices, and projections from these regions to the hippocampus, regulate working memory (Eriksson, Vogel, Lansner et al., 2015; Christophel, Klink, Spitzer et al., 2017). Like declarative memory in general, "researchers see working memory as an *emergent phenomenon* that arises from cooperation among many brain areas. As a result, the human brain is capable of multitasking and dealing with different kinds of information concurrently, switching from one task to another with great flexibility" (Corkin, 2013, p. 72).

One type of long-term declarative memory that is not always included in memory taxonomies is prospective memory. This consists in the ability to plan an action and remember to perform it at some point in the future. Prospective memory is critical for effective agency because many actions involve long-range planning. This requires that the subject hold the intention in mind from the time she forms it to the time she translates it into action. Prospective memory depends on stores of episodic and semantic memory. Holding and retrieving this information is necessary to form and complete action plans. Memory lapses associated with forgetting what one intends to do involve impairment in prospective memory (Einstein and McDaniel, 2005).

Prospective memory involves two neural pathways. A frontal-parietal pathway regulates attentional control processes in forming and maintaining an intention. This pathway also regulates how environmental cues indicate when the intention should be executed. These cues show that both temporal and spatial memory is necessary for diachronic agency. A second pathway in prospective memory involves the frontal-parietal cortex, hippocampus and anterior cingulate cortex (ACC). This pathway regulates retrieval of the memory and execution of the intention in action (McDaniel, Umanath, Einstein et al., 2015). By enabling the formation and execution of long-range intentions, the ACC has an important role in most aspects of agency. Dysfunction in any of these regions can impair the capacity to recall earlier intentions and result in failure to perform actions. This is another example of how memory systems can be highly interactive, depending on the nature and duration of the cognitive, emotional or motor task at hand. Intentions are action plans (Bratman, 2007). A voluntary action requires the ability to form and translate an intention into that action. Prospective memory enables one to anticipate performing an action and hold an intention to act until the appropriate time. Agency generally requires interaction between short- and long-term episodic, semantic, working and prospective memory.

Spatial memory is also necessary for agency. This refers to memories for locations of events and the events occurring at these locations. Spatial memory consists in the formation of environment-specific place maps. The information orients a subject in space and thus is critical for the contextualization of memories and actions that rely on it (Moser and Moser, 2008; Hasselmo, 2009). Hippocampal place cells and entorhinal grid cells enable spatial memory in the brain's navigational system (O'Keefe and Nadel, 1978; O'Keefe and Burgess, 1996; Moser and Moser, 2008, Hasselmo, 2009; Moser, Rowland and Moser, 2015). Grid cells are a more fundamental feature of this system and drive place cells. They form an internal positioning system, informing the organism of its location independently of external cues. Place cells use this information along with environmental cues to create a sense of space. They are sensitive to both internal sensory information in the brain and external sensory information from the environment with which the organism interacts.

The locus coeruleus in the brainstem has a critical role in this process. By sending inputs to the hippocampus, it enables the formation of neural representations of new places (Wagatsuma, Okuyama, Sun et al., 2018). The visual cortex is critical for the visual aspect of spatial memory, or visual-spatial memory, which depends on inputs from this part of the occipital lobe to the hippocampal-entorhinal circuit. Hippocampal networks that allow fast storage and retrieval of information about place may be at the core of declarative memory formation. Together with one's orientation to time, spatial memory provides one with the gist rather than details of the milieu in which one acts. One's orientation to space may be unconscious or conscious, depending on whether the cognitive demands of the milieu are habitual or novel.

Another subtype of episodic memory is recognition memory. As noted, this divides into familiarity and recollection involving independent retrieval mechanisms in a dual-process model (Mandler, 1980). Familiarity relies on cognitively fast, automatic processes. Recollection relies on cognitively slow reflective processes, requiring effort and attention in recognizing persons, words, objects or events (Eichenbaum, 2012, pp. 257–263). Retrieval of familiarity memory depends on processes in the perirhinal cortex, and retrieval of recollection memory depends on processes in the hippocampus. The division of labor between these two subtypes of recognition is another feature of the adaptive purpose of memory in allowing us to meet the cognitive demands of our environment. This distinction is relevant to the reliability of facial recognition and other general and specific features of describing a criminal act in legal testimony, which I discuss in Chapter 6. Familiarity can facilitate expeditious action in situations requiring rapid responses without being slowed by details. In situations requiring more complex cognitive processing, recollection can facilitate decision-making. Multiple memory systems are necessary for the mental events and bodily movements that constitute actions. Both types of recognition memory may slow depending on the amount of information the brain has to process and how efficiently it processes it. This occurs in aging brains with lower rates of neural firing and synaptic connectivity. This may explain why people recall a place or event only after some time and effort and with the help of external cues.

Nondeclarative memory includes procedural memory, priming, classical conditioning and nonassociative learning. Procedural memory involves the ability to perform motor skills such as driving a car or riding a bicycle. Following an initial period of conscious learning and information retention, procedural memory operates automatically. It is usually dissociable from episodic and semantic memory. This is illustrated in some cases of profound anterograde and retrograde amnesia, where people who cannot form new memories of events or facts or retrieve existing memories continue to perform motor skills.

In some cases, procedural memory and working memory may interact as part of a psychopathology. Patients with obsessive compulsive disorder (OCD) distrust procedural memory because of excessive rumination about ordinarily automatic actions like locking doors or turning off faucets (Van den Hout and Kindt, 2003). The obsessions and compulsions are symptomatic of distorted information processing that interferes with the ability to execute basic actions as a matter of course. Excessive reflection on simple cognitive and motor functions interferes with the rapid and efficient information processing necessary for working memory. It impairs reasoning and decision-making in those with OCD. This is a psychiatric disorder in which a dysfunction in (nondeclarative) procedural memory can cause dysfunction in (declarative) working memory.

Priming consists in exposing a subject to a stimulus that influences a response to a later stimulus without the subject being aware of it. It "involves initial presentation of a list of words, pictures of objects, or nonverbal materials and then subsequent re-exposure to fragments of items or very brief presentation of whole items" (Eichenbaum, 2012, p. 93; Tulving and Schacter, 1990). This type of nondeclarative memory may be intact in amnesic patients. Priming can have positive or negative effects on a person's behavior, depending on the initial presentation, the items that are presented and subsequent cues triggering a response to them. Priming may be associated with postoperative sequelae of implicit memory from surgery with or without conscious awareness under anesthesia. Classical conditioning involves the association of an external stimulus (the conditioned stimulus) with another stimulus (the unconditioned stimulus) to produce a reflexive unconditioned response. This results in the gradual acquisition of a conditioned response to the initial stimulus (Eichenbaum, 2012, pp. 384ff.). This process is pertinent to

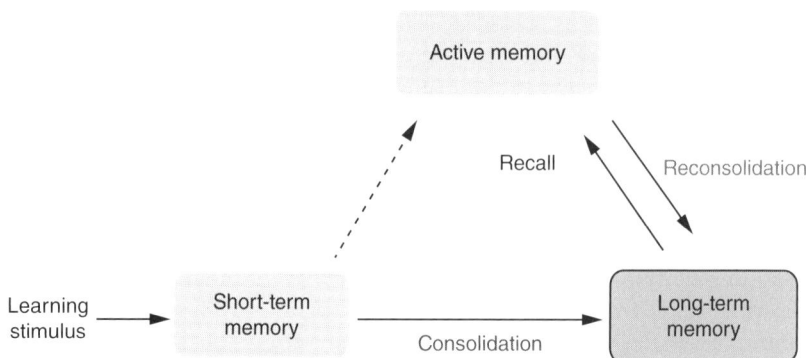

Figure 1.3 Memory consolidation and reconsolidation. The figure depicts the process of consolidating information encoded as short-term memory and retrieving and reconsolidating it as long-term memory. The two parallel solid arrows indicate interconnections between retrieval (recall) and reconsolidation when memories are destabilized, updated and restabilized. (with permission from Karen Rommelfanger, image creator)

attenuating emotionally charged memories in PTSD and anxiety through extinction training and other behavioral techniques.

The memory process consists in encoding, consolidating, storing, retrieving and reconsolidating information for further storage and retrieval (Thompson and Madigan, 2005, pp. 228–250; Eichenbaum, 2012, pp. 193ff.). Following a learning stimulus, the brain encodes information about it in a neural representation constituting the trace, or engram. The stimulus may be an experience leading to an episodic memory, or a word or fact leading to a semantic memory. Once the information is encoded, it goes into short-term storage. In normal brain function, memories are unstable before they become consolidated (McGaugh, 2000, 2015; LeDoux, 2007; Nadel, 2007). After they have been consolidated and transition from short-term to long-term memories, they again become unstable when they are retrieved. This later stage of instability is necessary for memories to become reconsolidated and restabilized to remain in long-term storage. The lability of the memory makes it susceptible to change. This destabilized state is what allows the subject and his brain to update the content of the memory so that it aligns with contextual information. There is a continuous causal sequence of neural and mental events, running from retrieval to reconsolidation, and back again. Updating occurs within this process (see Figure 1.3).

Contrary to historical hypotheses about long-term memory, contemporary researchers claim that "overwhelming evidence suggests that consolidated memories can be transferred again into a labile state, from which they are restabilized by a reconsolidation process. Retrieval appears to be the key process that transfers memory from the stable to the unstable state" (Nader, 2013, p. 3). It is during the labile state following retrieval when memories can be weakened or erased. The retrieval–reconsolidation process is consistent with Moscovitch's point that memory consolidation may never be complete. It is also consistent with Karim Nader's hypothesis that memory is a dynamic process and that memories need constant updating to remain relevant to the environment and promote adaptability to it (Nader, Schafe and LeDoux, 2000; Nader and Einarsson, 2010; Schiller and Phelps, 2011; Nader, 2013; Nader, Hardt, Einarsson et al., 2013; Lee, Nader and Schiller, 2017).

Memory systems are not fixed but flexible (Nadel, 2007; Nader, Hardt, Einarsson et al., 2013, p. 34) The adaptive function of memory, and what it enables us to adapt to, shows that it is not entirely a neurobiological process but one that depends on factors both inside and outside the brain. Memory is a product of interacting genetic, neurobiological, psychological and environmental factors. The content of our memories must reflect the actual natural and social conditions surrounding us. This is one reason why false or imagined episodic memories could not be adaptive, at least not over extended periods. They fail to reflect features of the external world and how we maintain or modify our behavior in responding to it.

The encoding, consolidating and storing of declarative memory requires a normal functioning hippocampus and adjacent structures in the hippocampal complex of the MTL. Episodic and semantic memories that have been encoded and consolidated then transition to frontal, temporal and parietal cortices for long-term storage and retrieval. "Connectivity between the prefrontal cortex, parietal cortex and the medial temporal lobe across different frequency bands underlies successful memory retrieval" (Knight and Eichenbaum, 2013, p. 257). According to MTT, episodic and semantic memories depend on different retrieval mechanisms. Semantic memories can be retrieved through cortical circuits without the hippocampus (Nadel and Moscovitch, 1997; Moscovitch and Nadel, 1998). But the hippocampus is necessary to retrieve episodic memories. Accessing these memories requires communication between hippocampal and cortical circuits. Once they have been encoded, consolidated and stored, semantic memories become independent of the hippocampal complex because they do not require recollection of the context in which we learned the information (Corkin, 2013, p. 223). The hippocampus is necessary to distinguish between contexts in which we experience events. Although episodic memories can become more like semantic memories over time, they still

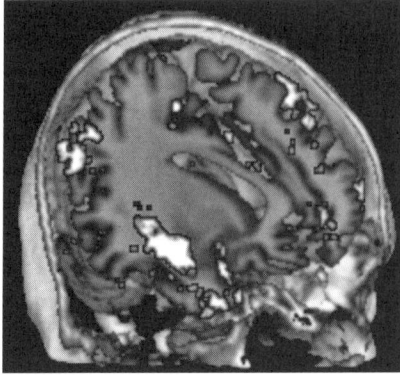

Figure 1.4 Brain activation in episodic memory retrieval. Automated meta-analysis of 332 functional MRI studies related to episodic memory, based on maps generated by neurosynth.org. The large white area in the lower center-left of the image is the hippocampus, and the smaller white areas above it are regions of the cortex. (with permission from Nick Davis)

require hippocampal function and hippocampal-cortical communication for retrieval (see Figure 1.4).

Emotional memories, including fear memories, have an affective representation in a limbic circuit that includes the amygdala and a cognitive representation in the neocortex (LeDoux, 2015). In addition to LTP and protein synthesis, emotionally charged memories of disturbing or traumatic events involve the action of the stress hormone norepinephrine in consolidating and storing them in the amygdala. Procedural memories rely on mechanisms in the subcortical striatum and cerebellum. Conditioning relies on mechanisms in limbic and subcortical regions. Priming relies on mechanisms in posterior regions of the neocortex. Like explicit memory, implicit memory involves a distributed neural network. Subcortical structures may project to cortical structures mediating cognitive and motor skills. The cerebellum regulates coordination, balance and other motor functions. In addition, it contributes to the ability to form action plans through its projections to the prefrontal cortex (Strick, Dum and Fiez, 2009). Among its nonmotor functions, the cerebellum has a direct role in both implicit and explicit memory processes. This and other subcortical structures may organize separate circuits for specialized types of memory (Tomlinson, Davis, Morgan et al., 2014a, 2014b). They may also interact with cortical structures in enabling interaction between procedural, episodic, semantic, working and prospective memory in agency.

Disorders of memory capacity involve dysregulation in one or more stages of the memory process. People affected by these disorders are unable to form, store, or access memories because of impaired mechanisms that ordinarily regulate this process. Anterograde amnesia is a disorder of memory consolidation. People affected by it cannot learn new information or form a stable representation of their experiences in new memories. Retrograde amnesia is a disorder of memory retrieval. People affected by it cannot recall information that has been consolidated and stored as memories in their brains. Because memories must be retrieved to be reconsolidated and remain in long-term storage, impaired retrieval mechanisms can result in the loss of these memories. Retrograde amnesia may also be a disorder of memory reconsolidation. Even when retrieval mechanisms are intact, and one can consciously access a memory, destabilization may allow extraneous information to intrude in this process and prevent reconsolidation and restabilization of the memory. It is also possible that protein synthesis and LTP have different effects on retrieval and reconsolidation. Retrieval alone may not be enough to ensure reconsolidation. These mechanisms could cause the memory trace to weaken and eventually disappear.

Dysfunction in retrieval or reconsolidation mechanisms, or both, may explain memory loss among some patients undergoing electroconvulsive therapy (ECT) for unipolar and bipolar depression. A plausible explanation for this phenomenon is that the induced seizures impair retrieval mechanisms in a hippocampal-cortical circuit. Temporary or permanent impairment in these mechanisms can cause temporary or permanent loss of episodic memories. Some patients with temporal lobe epilepsy (TLE) experience both anterograde and retrograde amnesia. Which type of amnesia they experience depends on how seizures affect consolidation (anterograde) and retrieval (retrograde) mechanisms (Zeman, Kapur and Jones-Gotman, 2012a). Although there may be problems with reconsolidation even when the capacity for retrieval is preserved, a problem with retrieval will eventually affect reconsolidation, since the first is necessary for the second. Brain damage or neurodegeneration causing chronic dysfunction in retrieval or reconsolidation mechanisms will gradually result in the loss of the information constituting the trace and remove it from storage in the brain. One could not retrieve a memory or restabilize it through reconsolidation because there would not be enough integrated information in the brain to sustain the engram.

Aphasia is a type of semantic retrograde amnesia. This is a unique form of amnesia because, unlike other forms of this disorder, it does not involve episodic memory. Antonio Damasio describes aphasia as "a disturbance of the comprehension and formulation of language caused by dysfunction in specific brain regions" (1992, p. 531). Those affected by this neurological disorder are unable to coherently express concepts and facts or comprehend them. They "forget" how to speak or understand language. The disturbance may be transient or progressive and may result from ischemia due to migraine, stroke, brain injury or neurodegenerative disease. These neurophysiological processes interfere with retrieval of information as stored words and simple and complex sentences. Aphasia is typically divided into two types, Broca's and Wernicke's aphasia, named after Paul Broca and Carl Wernicke (Broca, 1861; Wernicke, 1874; Damasio and Geschwind, 1984; Damasio, 1992). In the first type, also called nonfluent, or expressive, aphasia, affected persons have difficulty forming and completing coherent sentences. They lose the capacity for complex grammar from dysfunction in a region of the left frontal lobe. In the second type, also called fluent, or receptive, aphasia, affected persons have impaired speech comprehension from dysfunction in a region of the left temporal lobe. They speak fluently but cannot recall words and make them up as they speak (LaPointe and Stierwalt, 2018).

As part of the natural updating process, information associated with older memories may be removed from storage in the neocortex to assimilate information associated with newer memories. The provides contextual knowledge more relevant to one's present and future behavior. How much information is retained or replaced depends on how much information the brain can store and what is an optimal amount of information at a given time. This is another aspect of memory updating in consolidation and reconsolidation and how memories are transformed from short-term to long-term storage in the brain (Nadel, Hupbach, Gomez et al., 2012). Based on studies involving mouse models, in Alzheimer's disease, place cells, which are critical to spatial memory, fire out of sequence with other cells in the brain. This disrupts the encoding of new spatial memories in the degenerating hippocampus. The formation of new memories is also disrupted because older, contextually irrelevant memories are retrieved inappropriately. Encoding and retrieval mechanisms are dysfunctional and symptomatic of the disease (Cheng and Ji, 2013). Prefrontal degeneration and loss of hippocampal-cortical communication in Alzheimer's also impairs working memory.

At an advanced stage of the disease, all memory systems fail, including procedural memory. As I explain in Chapter 5, the potential of a hippocampal neural prosthetic to supplement or replace a degenerating hippocampus is significant and offers some hope for Alzheimer's patients. This is because the hippocampal-entorhinal circuit is one of the first regions of the brain to be affected by the amyloid-beta plaques and tau pathology resulting in neurofibrillary tangles in this disease. If clinical trials demonstrate its safety and efficacy, then this brain implant could restore some degree of spatial, working and other types of memory in patients with this and other neurodegenerative diseases.

In hyperthymesia, episodic memory consolidation and retrieval mechanisms are hyperactive (Parker, Cahill and McGaugh, 2006; Price, 2008). This condition correlates with increased volume of the parahippocampal gyrus, though researchers have not established a causal connection between this anatomical brain feature and the condition. Hyperthymesia has been described as "highly superior autobiographical memory (HSAM)" (LePort, Stark, McGaugh et al., 2015). Yet "hyper" suggests that it is a memory disorder rather than a form of enhanced memory. In hyperthymesia, recall is often involuntary, and the flood of information can impair some cognitive functions. In Chapter 2, I present cases illustrating how an excess of details in autobiographical or factual memory can interfere with the capacity for generalized inferential reasoning. HSAM may also impair working memory by distracting one from attending to immediate cognitive tasks. Excess episodic and semantic memory may impair the ability to learn new information and use it in planning for the future. In these respects, prodigious memory can limit rather than enhance control of thought and behavior.

Flexible behavior and adaptability to the environment require optimal levels of information in the brain and in memory retrieval between deficit and surfeit. Retrograde amnesia impairs control of thought and behavior due to a retrieval deficit. Hyperthymesia can also impair this control due to a retrieval surfeit. A study involving guinea pigs showed that hippocampal neurogenesis necessary for learning new information may also promote forgetting. This may be the brain's way of regulating information processing and its effects on behavior (Akers, Martinez-Canabal, Restivo et al., 2014). The results of this study are consistent with the view that, in normal brains, adaptive memory requires a balance between remembering and forgetting.

Disorders of memory content are disorders of the fear memory system. Unlike disorders of memory capacity, the problem is not an inability to form, store and access memories. Rather, the problem is an inability to remove memories that are maladaptive and pathological. It involves dysregulation at multiple stages of consolidation, storage, retrieval and reconsolidation. The emotionally charged content of the memory entrenches the representation of it in the amygdala. This causes the memory to persist beyond any adaptive purpose. A cognitive representation of the memory forms in the hippocampus and migrates to a storage site in the neocortex. An emotional representation of the memory forms in a hippocampal-amygdalar pathway and migrates to a storage site in the amygdala. Cues reminding the person of the experience on which the memory is based can trigger a repeated pattern of involuntary recall of the trace and its emotional content. The frequency of retrieval in response to cues can intensify reconsolidation and storage of the pathological memory, resulting in symptoms of generalized anxiety disorder (GAD) and PTSD.

In these disorders, many or all stages of memory formation and retention are on overdrive because of dysregulation in fear memory processing. Hyperactivation of this system is what makes the memory persist and refractory to many interventions. These memories constitute a distinct category from normal emotional memories because of their pathogenesis and adverse effects on thought and behavior. The difference between fear memories in GAD and PTSD is that the latter involves greater hyperactivation and persistence of the trace in the fear memory system.

Not all memory disorders are alike. Some involve cognitive impairment. Others involve both cognitive and emotional impairment. Still others involve motor impairment. Some involve a deficit of memory. Others involve a surfeit of memory or heightened emotional content. Different disorders result from dysfunction at one or more stages of the memory process. They can result from dysfunction in separate or interacting memory systems.

Understanding Memory from Brain Damage and Dysfunction

Research examining the behavior of patients with impaired memory from brain surgery, brain injury and neurological diseases has shed light on the neural structures mediating memory. It has elucidated how different neural structures mediate different types of memory. The research has shown that these structures are dissociable in regulating some memory functions and interactive in regulating others. The combination of behavioral observation, structural neuroimaging in the form of magnetic resonance imaging (MRI) or voxel-based morphometry (VBM), and functional neuroimaging in the form of positron emission tomography (PET) and functional magnetic resonance imaging (fMRI) have confirmed the neurobiological bases of memory.

In the 1930s and 1940s, Wilder Penfield used a metal probe with low-level electrical current to identify the source of seizures in awake patients on whom he was performing surgery for temporal lobe epilepsy. Although Penfield's principal aim was to reduce the incidence of the seizures, in performing this technique he produced maps of "functional localization" (Penfield and Boldrey, 1937; Penfield and Rasmussen, 1950; Penfield, 1975, pp. 21–27, 31–33). These maps helped him to locate the neural source of different experiences based on the patients' reports. In some patients, the stimulation elicited memories of past events, some of which were fairly accurate representations of them (Winter, 2012, pp. 75–93). Penfield's technique and his patients' reports indicated that structures in the MTL, particularly the hippocampus, were necessary for the encoding, consolidation and retrieval of episodic memories (Penfield, 1952).

More conclusive evidence of the role of the hippocampal complex in episodic memory came from studies of the famous amnesic patient "H.M." (Henry Molaison). These studies extend from the time William Scoville bilaterally removed his hippocampi and adjacent structures in his medial temporal lobes to stop his severe epileptic seizures in 1953 to just before his death in 2008. Scoville, Brenda Milner and Suzanne Corkin conducted tests on H.M. that have established the neurobiological foundation of memory and its wide-ranging effects on human thought and behavior (Milner and Penfield, 1955–1956; Scoville and Milner, 1957; Penfield and Milner, 1958; Milner, 1959; Squire, 2009; Corkin, 2013).

The temporal lobectomy left H.M. with profound anterograde amnesia. He was unable to learn new information from his experiences because he lacked the hippocampal structures necessary to form memories of them. His brain was only able to encode information for very brief periods, not long enough to stabilize it through consolidation. There were no new memories for him to store and retrieve. Researchers studying his behavior noted that "he could keep names, words or numbers in mind for a few seconds, and could repeat this information, but only when the memory load was small and nothing else intervened to wipe the slate clean" (Corkin, 2013, p. 63). He would have a conversation with another person and almost immediately forget that he had it. H.M. was unable to acquire episodic and semantic knowledge following the surgery.

This case contributed to the standard view of anterograde amnesia as the inability to form new memories from the time of injury onward. Yet generalized anterograde amnesia does not entirely exclude the ability to form new memories. In 1911, Edouard Claparede reported that a patient with amnesia from Korsakoff's syndrome was able to form some memories. One day, Claparede hid a pin in his hand that pricked the patient when they shook hands. From that day on, she refused to touch the doctor's hand. This case appeared to show that a patient can form new implicit memories despite the inability to form new explicit memories (Claparede, 1911; Nicolas, 1996). The extent of amnesia may depend on the cause and extent of brain dysfunction. While Claparede's case may support the view that amnesia is a disorder of declarative memory, the loss of the ability to learn new cognitive and motor skills due to advanced dementia or stroke suggests that a broad conception of anterograde amnesia includes both explicit and implicit memory.

H.M. could recall facts and concepts he had learned before the surgery. Except for a few emotionally significant experiences, though, he was unable to recall personal episodes from his past. While much of his semantic memory was intact, he had severe retrograde episodic amnesia as well. Corkin writes, "As we studied Henry over the years, we learned that his deficit was specific to his autobiographical memory – although he was unable to retrieve unique life experiences, he remained capable of recalling public events with considerable clarity ... When it came to personal information, however, his deficit was extreme ... Henry recalled facts but not experiences" (2013, p. 220). Because he lost his memory of them, he was unable to integrate his experiences into a personal narrative. Although H.M. could retrieve semantic memories from his presurgery past, he was unable to understand how they were connected and grasp their general meaning, their gist. Removing his hippocampi and adjacent structures in his MTL resulted in the loss of not only an autobiography but also the ability to construct a coherent account of the past.

There are two distinct senses of "gist" in the memory literature. One sense pertains to place and grid cells in the hippocampus and entorhinal cortex providing a person with the capacity for spatial orientation (Moser and Moser, 2008; Hasselmo, 2009, 2012). These cells allow a person to navigate the spatial landscape without having to recognize every feature of it. The other sense pertains to the meaning one constructs from information about the past as memories become less detailed and more generalized over time. These two senses of "gist" are complementary and involve interacting regulatory processes in the brain. In normal memory function, generalized spatial, episodic and semantic memory prevents the retention of unnecessary information about the past and promotes updating of information to make it relevant to one's surroundings. It is the brain's way of efficiently processing information and making memories adaptive for the organism and subject. In my discussion of declarative memory through the example of H.M., I have been using "gist" mainly in the second sense. This was just one aspect of declarative memory that he lost.

The effects of the surgery on H.M.'s memory confirm a key feature of MTT. Whereas retrieval of semantic information does not require the hippocampal complex, retrieval of episodic information does require it. Retrieval of semantic memories requires only frontal, temporal and parietal cortices. Retrieval of episodic memories requires these cortical regions and regions of the hippocampal complex. "According to the Multiple Trace Theory, it is possible to retrieve facts – like 1492 – without the hippocampal system, but it is not possible to retrieve any unique experiences – like [a] birthday celebration – unless the hippo-campal circuits can communicate with the cortical circuits" (Corkin, 2013, p. 224). The few episodic memories H.M. retained from his presurgery experience had salience and emotional significance. This made the memories robust and "available for retrieval decades later" (p. 228). It may be inaccurate the describe his retrograde amnesia as temporally graded relative to his surgery because he retained so few episodic memories of experiences before the surgery. Nevertheless, his ability to retrieve some of these memories suggests that the surgery did not completely destroy his hippocampal complex. Corkin points out that a segment of the dorsal region of H.M.'s parahippocampal gyrus was spared (p. 228). The fact that the preserved episodic memories were of emotionally arousing experiences suggests that sections of other limbic areas like the amygdala, and connections between the remaining section of the hippocampus and the amygdala, were also spared.

This may also explain why another person with profound amnesia, musicologist Clive Wearing, is able to gradually recognize his wife after an initial period of unfamiliarity each time he sees her (Sacks, 2007). Although herpes viral encephalitis when he was in his mid-forties (he is now eighty-one) destroyed many of the brain regions mediating declara-tive memory, sections of the hippocampal complex and the amygdala were probably spared. This has enabled him to retain and retrieve some memories of emotionally arousing experiences associated with his rela-tionship with his wife, albeit with difficulty and only to a limited extent. In addition, the fact that he continues playing the piano indicates that at least some of his procedural memory is intact.

The effects of H.M.'s retrograde amnesia from the loss of his hippo-campi confirm Bartlett's reconstructive memory theory and more recent formulations of this theory by contemporary memory researchers. Con-sistent with Schacter and Addis's position that MTL-mediated episodic memory enables us to imagine the future, Corkin writes: "The process of imagining future events depends on medial temporal-lobe structures, the prefrontal cortex and the posterior parietal cortex – the same areas critical for declarative memory. Remembering … past events and recom-bining them to create future scenarios requires the retrieval of infor-mation from long-term memory, and it is no surprise that amnesia interferes with this process. Constructing the future, like resurrecting the past, requires establishing functional connections between the hippocampus and areas in the frontal, cingulate and parietal cortices. Without this network, Henry had no database to consult when asked what he would do the next day, week or in the years to come. He could not imagine the future any more than he could remember the past" (Hassabis and Maguire, 2007; Hassabis, Kumaran, Vann et al., 2007; Addis and Schacter, 2012; Corkin, 2013, pp. 235–236).

This assessment of H.M. is also consistent with the idea that a core set of episodic and semantic memories of actual events and facts provides an informational scaffolding necessary for goal-directed behavior. His retrograde amnesia caused him to lose this scaffolding and in turn his prospective memory. It precluded the capacity for any sort of planning. "Henry's brain could not translate information from short-term processing mechanisms to long-term processing mechanisms" (Corkin, 2013, p. 71). His anterograde amnesia prevented his brain from consolidating new information. His retrograde amnesia prevented his brain from retrieving stored information and reconsolidating this information. Both forms of amnesia precluded him from forming, sustaining and translating complex intentions into actions.

In addition, H.M.'s case sheds light on the differences between familiarity and recollection and how they are mediated by different neural networks. He displayed familiarity by identifying magazine pictures and other objects he had seen before. But he had no recollection of his personal experience. This requires more cognitive effort in concentration and attention in making sense of experiences by contextualizing them and understanding how they are connected. The combination of H.M.'s behavior and post-surgery structural imaging led to the knowledge that familiarity is enabled by the perirhinal cortex and that recollection is enabled by the parahippocampal cortex. Michael Rugg and Kaia Vilberg explain, "Recollection-related activity in the parahippocampal cortex has a central role in the representation of contextual information, retrieval of which is a defining feature of successful recollection" (Rugg and Vilberg, 2013, p. 255). H.M. lacked this capacity because of his cognitive inability to connect information from different sources across time. As part of the adaptive function of declarative memory, a memory that is retrieved and reconsolidated is one component of an integrated diachronic informational network rather than an isolated fragment of experience. New content must be assimilated into or replace existing content to allow a person to meet her actual cognitive, emotional and motor needs. Dysfunctional retrieval and/or reconsolidation mechanisms in retrograde amnesia disrupt this process.

Although H.M.'s anterograde and retrograde amnesia severely impaired his agency, it did not eliminate it. His procedural memory was preserved. He retained his knowledge of how to perform motor tasks. Beyond an initial learning period, these skills do not require conscious processes involving connections between hippocampal and cortical networks. The subcortical striatum and cerebellum that mediated his procedural memory were intact. Unlike declarative memory, which may take years to become stable, nondeclarative systems can establish procedural memory of motor functions in a much shorter period. But H.M.'s procedural memory deteriorated over many years because chronic use of the antiseizure drug phenytoin (Dilantin) caused gradual degeneration of his cerebellum. Just as the correlation between his conscious behavior and his postsurgery brain confirmed the limbic and cortical foundation of declarative memory, the correlation between his decreasing motor capacity and gradual damage to the cerebellum confirmed the subcortical foundation of procedural memory. H.M.'s immediate declarative memory impairment resulted from the bilateral medial temporal lobectomy. His gradual nondeclarative memory impairment resulted from the effects of Dilantin.

Because working memory largely depends on the prefrontal cortex, and because this region of H.M.'s brain was intact, he also retained some capacity for working memory. He was able to sustain attention and hold and use information long enough to successfully perform some card-sorting tasks. Yet even his cognitive ability to perform simple tasks of this sort was limited. He retained access to facts and concepts and was able to use them to some extent in some basic forms of problem-solving. Still, the disrupted connection between his missing hippocampal complex and the prefrontal cortex and the loss of his episodic memory severely limited the amount of information available for him to make complex decisions. Although he retained some procedural, working and fragmented semantic memory, H.M. was severely deficient in his agency in the respects I have described. I will return to his case and consider other cases of amnesia and their implications for agency in Chapter 2.

Recent research has refined our understanding of episodic memory retrieval mechanisms in the brain. Studies of H.M. and other amnesic patients indicate that a circuit consisting of the hippocampus and entorhinal cortex is necessary for both consolidation and retrieval of episodic memories. Based on a study using mice, a group of neuroscientists has shown that a circuit involved in retrieval is separate from the circuit involved in encoding and consolidating these memories (Kitamura, Ogawa, Roy et al., 2017; Roy, Kitamura, Okuyama et al., 2017). The additional structure in the more complex retrieval circuit is the subiculum. This structure is part of the hippocampal complex located between the hippocampus and entorhinal cortex. Although one must be cautious in drawing inferences from results of animal studies to potential effects in humans, this study suggests that patients with an intact hippocampal-entorhinal circuit who can form, consolidate and store memories may not be able to recall them if the subiculum does not connect or communicate with the hippocampus and entorhinal cortex. Dysfunction in the subiculum may account for problems with episodic memory retrieval in retrograde amnesia. The trace may be in long-term storage but not accessible for recall. One possible explanation for why only one circuit is necessary for memory formation but two circuits are necessary for memory retrieval is that interaction between the hippocampal-entorhinal and hippocampal-subiculum-entorhinal circuit facilitates memory editing and updating. This interaction may be necessary to allow new information to be added to or replace existing information in the brain.

An intriguing aspect of this dual-circuit model is that it could be part of an explanation for the pathogenesis of Alzheimer's disease (AD). At an early stage, patients with this disease have difficulty recalling events but may retain some capacity to form new memories and learn new information. The more complex hippocampal-subiculum-entorhinal circuit may be affected by tau pathology to a greater degree than the hippocampal-entorhinal circuit and explain why patients with AD are unable to retrieve spatial and episodic memories. Reconsolidation is necessary for memories to be restabilized and remain in long-term storage. This follows memory retrieval, when memories become destabilized. Chronic dysfunction in retrieval will result in chronic dysfunction in reconsolidation and loss of the memory trace. Dysfunctional retrieval mechanisms may explain memory loss in other dementias as well.

Patients with advanced AD also have impaired capacity to form new memories. Research suggests that this may result from reduced production of dopamine by the ventral tegmental area for the hippocampus (De Marco and Venneri, 2018). By a certain stage of the disease, Alzheimer's patients are impaired in both retrieving existing memories and consolidating new memories. The loss of the brain's information consolidation function eventually results in the loss of information available for retrieval.

Conclusion

Memory is a process through which information about the past is encoded, consolidated and stored in the brain as a representation of our experience. It is not merely a storehouse of ideas that we reproduce when we recall events. Memory is a database or scaffolding of information from which we constantly draw in simulating or imagining possibilities and projecting ourselves into the future. It has both backward-looking and forward-looking dimensions. Memories are not entities identical to neural nuclei or tissue. They cannot be visualized in discrete regions of the brain. Instead, they correspond to more or less integrated levels of information processing. Memory is a neurobiological process in which representations of experiences emerge from and are sustained by neural firing, synaptic connectivity and other mechanisms in separate and interacting neural networks. Memory is also a psychological process in which the cognitive and emotional content of experience is captured in a person's conscious and unconscious mental states and displayed in her behavior. The fact that episodic memories become less detailed and more general over time is an indication of their adaptive purpose. This allows the brain to remove irrelevant information and assimilate new information that is more relevant to one's present and future circumstances. It allows memories to be updated so that they promote flexible behavior and adaptability to the environment.

There are optimal levels of memory in normal brain function. Too many or too few episodic and semantic memories can inhibit flexibility and adaptability. The destabilization of memories that occurs when they are retrieved enables them to be updated before they are restabilized and re-stored in the brain through reconsolidation. These processes are features of the constructive and reconstructive model of declarative memory. Still, this model does not provide a satisfactory account of working memory. Processing and using information for immediate cognitive tasks does not appear to involve updating. Nor does it explain how nondeclarative procedural memory also has an adaptive purpose. The cognitive and motor skills in this type of memory do not appear to involve updating either. Constructive and reconstructive models are thus limited in providing a comprehensive explanation for all memory systems.

The neuroethical issues surrounding disorders of memory capacity and content, and interventions to treat them, arise from how they affect our experience and influence normative judgments about our behavior. These issues also arise from interventions designed to enhance normal memory functions. In the chapters that follow, after explaining the role of memory in agency and identity, I analyze and discuss how disorders of memory content and memory capacity affect people's experience of mental time travel and capacity for goal-directed behavior. I explore potential benefits and risks of psychotropic drugs and neuromodulation used to weaken or erase pathological and disturbing memories. In addition, I discuss the potential benefits and risks of memory-modifying drugs and techniques used to improve, restore or enhance memory functions. And I examine legal issues pertinent to judgments of memory capacity and different forms of memory modification. My general discussion highlights the critical role of memory in our lives and the normative implications of measuring, modifying and erasing it.

2 Agency, Identity and Dementia

Memory is necessary to form, hold and execute intentions in actions. Depending on whether these plans pertain to present or future projects, how we develop and translate them into actions depends on different memory systems. A robust sense of agency in goal-directed behavior requires proper function of episodic, semantic, working and prospective memory. Physical actions in the form of bodily movements also require procedural memory. Both declarative and nondeclarative memory are necessary to perform basic and complex actions. Impairment in one or more of these systems can impair agency to varying degrees. This depends on the cognitive or motor task at hand and which memory systems are critical for these tasks.

Memory is also necessary for personal identity. The connections between mental states that enable the experience of persisting through time as the same individual require the ability to recall experiences from the past and use them to project oneself into the future. Amnesia mainly affects episodic memory. Disorders of episodic memory consolidation, storage, retrieval and reconsolidation can disrupt the psychological relations necessary for identity and the experience of mental time travel. In advanced dementia, there is substantial weakening or loss of these psychological relations and the ability to recall the past or imagine the future. One may no longer be the same person or an agent because these relations no longer hold. In the last stage of dementia, there is loss of awareness and a vanishing of the self.

In this chapter, I examine how memory function enables agency and how memory dysfunction disables agency. Because some memory systems mediating goal-directed behavior may be intact while others are dysfunctional, neurological and psychiatric disorders that impair some of these systems can impair agency to varying degrees. Yet some people with severe amnesia may retain some cognitive and motor functions and some degree of agency. I present actual and hypothetical cases to illustrate these points. In addition, I analyze and discuss the role of memory in personal identity. Specifically, I consider how much episodic memory is necessary to meet the transitivity condition of identity and how much of this memory can be weakened or lost without violating this condition. These considerations are motivated by the idea that memory is not merely a reproduction but a reconstruction of the past that allows us to simulate future events and influence them through our actions. Reconstructing memory prevents the brain from being cluttered with trivial details about the past. But a core set of episodic memories is necessary to form a constant and stable cognitive and emotional scaffolding for identity and forward-looking thought and behavior.

Updating episodic memories to adapt to current and future circumstances can alter the content of these memories. It can weaken the connections between one's present and past mental states. Adaptability may come at the cost of identity in a radically extended lifespan. In the last section of the chapter, I discuss how the gradual loss of declarative memory in dementia influences judgments about wishes expressed by a competent person at an earlier time about how he should be treated at a later time. This involves the concept of precedent autonomy. If the demented individual is not the same person as the individual who was cognitively intact, then should caregivers respect and act on an earlier directive given the change in mental status and identity? Does the patient's autonomy in expressing his wishes about future medical treatment transfer from a competent state to a demented state? While a person with late-stage dementia loses memory and all but experiential interests, a person with early-stage dementia may retain enough working and prospective memory to sustain a critical interest in how his remaining life should go. This may involve a genuine change in mind and interests from those he held earlier. When these interests conflict, the question arises as to whether caregivers should act in accord with the patient's earlier or later interests. All of these issues show how agency and identity are interconnected through the mechanisms of memory. I discuss agency before identity because one can act consciously or unconsciously using memory at distinct times without meeting the criteria of personal identity in terms of psychological relations holding over time. Agency can be a synchronic or diachronic phenomenon. Identity is a diachronic phenomenon.

Memory and Agency

Agency includes mental acts such as decisions and physical acts as bodily movements (Davidson, 2001). Actions are autonomous when they are performed voluntarily from considered desires and reasons the person endorses as her own. Agency involves psychomotor, cognitive, emotional and volitional functions, and memory has a critical role in all of them. Depending on the type of action one performs, this may involve episodic, semantic, working and prospective memory. Indeed, many actions require the coordinated function of all forms of declarative memory. They also depend on nondeclarative procedural memory. Although memory systems are dissociable, they often interact in many mental and physical capacities. Dysfunction in one system may cause dysfunction in others and disrupt voluntary agency.

According to Baddeley, working memory consists of three components: the central executive, the phonological loop and the visuospatial sketchpad (2007, p. 7). I leave aside the technical details in Baddeley's account and focus on its main features. "Working memory … sits at the interface between perception and action, between learning and attention, substantially increasing the flexibility of the organism" (p. 338). Further, "working memory storage capacity is small." It steps in "whenever this hugely varied and complex set of automatic processes appears to be failing, either through informational overload, through the demand for novel processes, or for future planning" (p. 338). This passage suggests that declarative working memory does not operate independently of nondeclarative procedural memory and the automatic processes it regulates. Working memory is a conscious process activated when procedural memory cannot meet the cognitive demands of the environment on the subject.

In Chapter 1, I described obsessive-compulsive disorder as distrust of procedural memory. One loses confidence in the ability to automatically perform basic motor and cognitive functions and engages in repetitive behaviors as a maladaptive attempt at reassurance. Rather than step in when automatic processes fail, in OCD excessive rumination causes hyperactivation of working memory and impairs one's ability to perform basic motor and cognitive functions as a matter of course. Because of dysfunctional connectivity in a prefrontal thalamic-striatal-prefrontal circuit, the subject misinterprets information in his semantic and episodic memory store, which in turn impairs his working memory. The obsessions and compulsions impair reasoning and decision-making.

When there is normal communication between the relevant brain regions, working memory involves the efficient information processing and attention necessary for less and more complex forms of reasoning and decision-making (Rolls, 2007). Chess masters and champions of memory competitions are highly proficient in working memory. In contrast, executive dysfunction in people with OCD, schizophrenia and dementia is associated with impaired working memory (Insel, 2010). Those with the positive subtype of schizophrenia characterized by hallucinations and delusions have difficulty filtering information in the brain.

The information overload and distortion undermine their ability to sustain attention and hold the information long enough to make rational decisions. A similar problem besets those with Alzheimer's disease, and the impairment increases as the disease progresses. They have difficulty learning and applying new information, as well as difficulty retrieving stored episodic and semantic memories in reasoning and planning. In neuropsychiatric disorders, dysfunctional working memory impairs agency primarily by interfering with the mental process of decision-making.

Agency in Alzheimer's patients is even more limited by the loss of hippocampal place cells and entorhinal grid cells from tau protein pathology (Fu, Rodriguez, Herman et al., 2017). These cells regulate the spatial memory that enables orientation to one's surroundings. They constitute the brain's "GPS" (Nielson, Smith, Sreekumar et al., 2015). Spatial memory grounds the capacity for planning necessary to navigate the physical world (Buzsaki and Moser, 2013). Deterioration of this type of memory causes spatial and temporal disorientation and loss of understanding of the context in which one lives and acts. The integration of spatial, working and episodic memory grounds the experience of persisting through time (Eichenbaum, 2012, pp. 219ff.; 2017). Difficulty in spatial navigation may be the first sign of cognitive decline in Alzheimer's (Allison, Fagan, Morris et al., 2016). Spatial and working memory are typically affected first by this disease because of lost place and grid cells and degeneration of dopaminergic neurons in hippocampal-entorhinal and prefrontal-hippocampal circuits.

Aphasia is another neurological disorder that can impede agency. In Chapter 1, I described Broca's and Wernicke's aphasia as types of amnesia involving semantic memory. In Broca's (nonfluent) aphasia, one forgets how to speak grammatically and coherently. In Wernecke's (fluent) aphasia, one can speak but forgets how to comprehend the language one uses or hears from others. They are both memory and communication disorders. They limit social interaction by impairing the affected person's ability to coherently report her experience or verbally respond to others. The loss of this ability may be temporary in cases of migraine or transient ischemic attacks. It usually resolves with restoration of normal blood flow in the critical brain regions. However, the loss may be permanent in cases of progressive aphasia due to a neurodegenerative disorder (LaPointe and Stierwalt, 2018).

Additional comments by Baddeley point to a link between working memory and the ideas of simulation and imagination in constructive accounts of episodic and semantic memory. "If we are to progress from simple reaction to planful action, then we need a device for planning and a system for holding and manipulating our plans. Planning essentially involves predicting what will happen. An obvious way to do this is to simulate an occasion that has occurred in the past. Such simulation requires a storage system with links to both perception and LTM (long-term memory) – in short, a working memory" (2007, p. 338). Working memory is a necessary intermediary that enables us to use information from the past in planning for the future. Baddeley's comment about holding action plans suggests a connection between working and prospective memory.

Prospective memory is necessary to sustain an intention in an ongoing or future task. As I explained in Chapter 1, prospective memory allows one to form and hold an intention about a future action until the time when one executes the intention in the action. Event-based prospective memory involves remembering to perform an action when a specific event occurs. Time-based prospective memory involves remembering to perform an action at a particular time. Both types of prospective memory depend on triggering or target cues in the environment that enable one to hold the intention and execute it in the appropriate place and at the appropriate time (McDaniel, Umanath, Epstein et al., 2015). The duration of this process may be considerably longer than working memory, which typically lasts for only a few seconds. Prospective memory may involve minutes, hours, days or even years. It requires a certain store of retrospective episodic and semantic memory as a basis on which to initiate and complete temporally extended action plans.

The rate at which items are recalled in retrospective memory is roughly the same in prospective memory. This implies that the same neurobiological mechanism underlies both types (Maylor, Chater and Brown, 2001). They contribute to the connection between agency and personal identity, enabling us to be aware of ourselves as subjects existing and acting through time. The fact that a prefrontal-hippocampal circuit mediates both working and prospective memory shows how closely connected they are in enabling synchronic and diachronic decision-making (McDaniel, Umanath, Epstein et al., 2015). An entorhinal-hippocampal circuit bridges temporal gaps between discontinuous events and is also necessary for prospective memory (Kitamura, Macdonald and Tonegawa, 2015). Dysfunction in either, or both, of these circuits can impair intentional agency.

Consider a parent who leaves a child unattended in a car but has an intention to return to the car within a brief period. This involves an episodic memory of leaving the child in the car, an intention to return to the car and a memory of forming that intention. By returning to the car and the child within this period, the parent demonstrates that she held the intention throughout this period and translated it into the appropriate action. She successfully demonstrated prospective memory. However, if she failed to return to the car within the critical period, then she failed to hold and act on the intention and thus demonstrated prospective memory failure. This omission may be symptomatic of a more basic failure of memory encoding or retrieval. It can have significant legal implications regarding criminal negligence causing harm or death. I discuss such a case in Chapter 6. These cases can be difficult to adjudicate because it may not be clear whether the memory lapse is psychogenic and within the agent's control or due to neurobiological dysfunction beyond her control. A brain abnormality alone may not provide a satisfactory explanation for failing to hold, retrieve and act on a prospective memory

Suppose that a researcher is giving a 30-minute presentation on a particular topic. In the introduction, he says that he will discuss a controversial hypothesis toward the end of the presentation. Discussing it would demonstrate successful exercise of prospective semantic memory regarding facts and concepts. His ability to execute his intention to discuss the issue requires forming and retaining a memory of the intention. If he failed to discuss the issue after announcing that he would, then his prospective memory would have failed. He would have forgotten his intention. Like the earlier case of the parent leaving the child unattended in the car, the failure may be due to a psychological lapse or an abnormality in a neural circuit mediating semantic memory encoding or retrieval. However, a correlation between the brain abnormality and the memory lapse would not necessarily mean that the first caused the second.

These cases illustrate how one can have generally normal working, episodic and semantic memory but impaired prospective memory. Despite forming an intention to act within a certain period, one may forget the intention and fail to act. One may otherwise display a significant degree of effective agency in performing many cognitive tasks. But the temporal extent of one's agency may be limited by a reduced capacity for prospective memory.

The capacity for effective agency presupposes optimal levels of episodic and semantic memory. The use of "highly superior autobiographical memory" to describe hyperthymesia might suggest that people with this condition have enhanced memory capacity. It might suggest that a high volume of memory storage and retrieval makes one's behavior more flexible and adaptive. But this is not necessarily the case. On the contrary, exceptional autobiographical memory may impede agency and be maladaptive by interfering with working and prospective memory. In her book *The Woman Who Can't Forget*, Jill Price describes how from age fourteen she noticed that she could recall events she had experienced in almost perfect detail. Although her capacity for episodic memory retrieval benefited her in her work as an administrative assistant at a law firm, most of her memories were filled with trivial and intrusive details. Price writes, "I don't make any effort to call memories up; they just fill my mind. In fact, they're not under my conscious control, and as much as I'd like to, I can't stop them" (Price, 2008, p. 135). She describes her exceptional autobiographical memory as "a horrible distraction" (p. 135). Further remarks by Price underscore the importance of constructive memory in eliminating unnecessary information about the past. "I've come to understand that there is a real value in being able to forget a good deal about our lives" (p. 43). "Instead, I remember all the clutter" (p. 45). This clutter of information can interfere with the ability to plan and act.

Cases of exceptional semantic memory also inhibit rather than enhance agency. An excess of facts about the past can interfere with the ability to use this information to imagine and simulate future events. It can interfere with the ability to use what one learns as a basis for planning and decision-making. The Russian neuropsychologist A. R. Luria describes his patient Solomon Shereshevskii as someone who had a seemingly limitless capacity to recall facts and numbers (Luria, 1969). This was associated with his synesthesia, which "gave his memories a richer content and made them easier to recollect" (Quiroga, 2012, p. 19). This actual case lies at the other extreme of the memory spectrum from the semantic amnesia of Macondo's fictional inhabitants in *One Hundred Years of Solitude*. Despite Shereshevskii's exceptional ability to recall and recite facts, he was unable to hold any line of work other than a traveling mnemonist. He seemed unable to capture the gist of his semantic knowledge because his mind was replete with numbers and letters.

Consistent with a point made by Schacter and Addis, Shereskevskii's inability to understand the general meaning of what he learned and use this knowledge to think about the future resulted from his mind being cluttered with trivial details that were not relevant to the cognitive demands of his natural and social milieu. Shereshevskii "had to make a voluntary effort to achieve what is innate in everyone else: to forget details and generalize" (Quiroga, 2012, p. 47). His problem was not an inability to remember facts, but an inability to forget them. Like Jill Price's autobiographical memories, Shereshevskii's semantic memories became a burden, a "torment" for him (p. 46).

There are similarities between Shereshevskii and the fictional character Funes in Jorge Luis Borges's short story,"Funes, the Memorious" (Borges, 1962, pp. 107–116) Following an equestrian accident in which he sustains a brain injury, Funes remembers every detail of everything he experiences. Unlike Shereshevskii, Funes's capacity for total recall is not associated with synesthesia. Rather, it could be explained by hyperactive consolidation, storage and retrieval mechanisms in his brain. Consistent with Borgesian irony, a brain injury that should have resulted in amnesia had the opposite effect of exceptional storage and recall of information. Speaking to the narrator, Funes says, "my memory, sir, is like a garbage disposal" (p. 114). He becomes an invalid, a prisoner of excess episodic and semantic memory. He is unable to learn anything, to conceptualize or to imagine future possibilities because there is no room in his brain and mind for any new information. The narrator says that Funes "was not very capable of thought. To think is to forget a difference, to generalize, to abstract. In the overly replete world of Funes, there were nothing but details, almost contiguous details" (p. 114). Funes's overloaded episodic and semantic memory impairs his working and prospective memory. It impairs the capacity to draw inferences from past events to possible future events and to form and execute action plans. Borges writes that Funes died of "pulmonary congestion" (p. 115), a metaphor for his overloaded memory.

Exceptional declarative memory may enable some people to perform certain cognitive tasks exceptionally well. Beyond a certain level, though, storing and retrieving a large volume of episodic or semantic memory can interfere with working memory by impairing efficient information processing necessary for executive tasks. In addition, it can cause one to focus too much on the past and thereby impair prospective memory. Excess memory may also interfere with the capacity for abstract thinking and to simulate future events.

Just because a certain amount of memory enables one to perform well on certain tasks, it does not follow that more memory enables one to perform better on these or other tasks. The "horrible distraction" of Jill Price's autobiographical memory and the "torment" of Solomon Shereshevskii's semantic memory are in stark contrast to Pliny the Elder's belief that it was a blessing to possess extraordinary memory (Pliny, 1942, pp. 562–563, cited by Quiroga, 2012, p. 15). Price's and Shereshevskii's conditions support the view that there are limits to the volume of memory that can be adaptive. They point to Nietzsche's claim that "all action requires forgetting, just as the existence of all organic things requires not only light but darkness as well" (Nietzsche, 1872–1874/1995, p. 65; cited by Quiroga, 2012, p. 20). The amnesia of H.M. illustrates how too little episodic memory can weaken or undermine the capacity to form and execute action plans over shorter or longer periods. The hyperthymesia of Jill Price and Solomon Shereshevskii illustrates how too much episodic and semantic memory can also weaken or undermine this capacity by causing information overload in the brain and mind.

An excess or deficit of episodic or semantic memory can be maladaptive by interfering with working and prospective memory. Optimal levels of remembering and forgetting are necessary to learn new information and use it to respond appropriately to one's circumstances. It is also important to emphasize that the information retrieved by people with hyperthymesia may include many false memories (Patihis, Frenda, LePort et al., 2013). If a higher rate of memory retrieval correlated with a higher rate of false memories, and if the misinformation in these memories impaired decision-making, then this would be another respect in which hyperthymesia impaired agency.

People with neurological diseases impairing subtypes of declarative memory may retain a limited degree of agency through intact procedural memory. For example, Clive Wearing's severe anterograde and retrograde amnesia prevented him from learning new information and applying what he had learned before the herpes viral encephalitis to carry out cognitive tasks. Yet he was able to continue performing on the piano. He retained his knowledge of how to play this musical instrument because his striatum and cerebellum were likely spared from the infection. This can also be explained by the emotional significance of music for him and what was likely an intact emotional representation of musical performance in his amygdala.

Even some patients with general impairment in motor functions from neurodegenerative diseases may retain the ability to execute certain motor skills. They can unconsciously remember how to perform certain actions while forgetting how to perform others. In a public discussion of movement disorders, broadcaster and former race-car diver Sam Posey commented that, despite the general motor impairment he experienced from Parkinson's disease, he was still able to drive a car without much difficulty (Rose, 2012). His intact procedural memory for driving could be attributed to the major role that this repeated activity had over many years of his life. The representation of this procedural memory had become so embedded in the motor circuit of his basal ganglia and cerebellum that it seemed immune to the degeneration of dopaminergic neurons in his brain. Like Clive Wearing's piano playing, the repeated recruitment of motor circuits and the emotional representation associated with driving can enhance this motor skill and keep it intact until the most advanced stage of the disease.

The case of jazz guitarist Pat Martino is another example of how preservation of some degree of procedural memory can enable some degree of agency when brain regions mediating declarative memory have been extensively damaged. After a ruptured cerebral aneurysm in 1980 left him with profound amnesia, Martino had to relearn many basic tasks. He also relearned how to play the guitar and recently resumed playing in public. Although the hemorrhage affected his procedural memory, repeated playing of the guitar over many years may have allowed some of this memory to resist and survive the injury. This may have facilitated relearning this activity (School, 2014). Nevertheless, without episodic, semantic, working and prospective memory, a person with intact procedural memory could perform only a limited number of tasks and would be significantly impaired in her agency. The behavior of Clive Wearing and H.M. are two examples.

I have noted that H.M.'s inability to recall information from the past and encode new information made him unable to imagine and plan for even the very near future. His depleted store of episodic and semantic memories severely limited his agency to cognitive and motor tasks involving mainly procedural memory. His inability to convert short-term memories into enduring ones caused him to be "stuck" in a permanent present tense (Corkin, 2013, p. 75), Oliver Sacks makes a similar observation of Clive Wearing: "To imagine the future was no more possible for Clive than to remember the past – both were engulfed by the onslaught of amnesia" (Sacks, 2007, p. 101). Other researchers have made similar claims that people with profound amnesia are unable to simulate future possibilities necessary for deliberation and decision-making (Hassabis, Kumaran, Vann et al., 2007). Presumably, they lack the capacity for constructive memory. However, other cases of amnesiacs challenge these claims. While in some cases amnesia severely limits agency, in others an affected person's ability to experience duration may be preserved. This may allow some degree of agency despite memory loss.

The patient "K.C." (Kent Cochrane) had severe anterograde amnesia and temporally graded retrograde amnesia from a traumatic brain injury following a motorcycle accident in 1981 (Tulving, 1985a). Despite the memory impairment from damage to his hippocampal complex, he retained various temporal conscious capacities (Craver, Kwan, Steindam et al., 2014). Still, he lost the capacity for episodic future simulation and to imagine his personal future. This case confirms that goal-directed thought and behavior require episodic memory. But it also suggests that not all severe amnesiacs lack an understanding of time (Squire, van der Horst, McDuff et al., 2010).

One possible explanation for the differences between H.M. and K.C. in his preserved temporal capacities is that some of K.C.'s hippocampal place cells and entorhinal grid cells and their projections to cortical networks were preserved and not affected by the brain injury. Or these cells were not damaged to the same extent as they were in H.M. The insular cortex, which regulates feeling states of the body, emotions and the processing of duration, may have been intact as well (Wittmann, 2013). K.C.'s amnesia might not have been of the same degree as that of H.M. or Clive Wearing.

Not all cases of anterograde and retrograde amnesia are the same. Impaired episodic memory does not always entail complete loss of the ability to understand time and a limited sense of connections between actions and their consequences (Irish, Addis, Hodges et al., 2012). The differences between these cases likely reflect differences in damage to the hippocampal complex, how this damage affects its projections to cortical networks and the extent of amnesia resulting from it.

Another disorder that appears to refute claims that extensive memory loss precludes intentional agency is severely deficient autobiographical memory (SDAM). People affected by SDAM have impaired episodic memory retrieval mechanisms. Yet they can continue to learn, retain and use information for certain tasks by recruiting nonepisodic processes (Palombo, Alain, Soderlund et al., 2015). They are able to perform these tasks even in the absence of the capacity for recollection. Presumably, familiarity and semantic knowledge would be sufficient for them to learn and apply information. Still, the absence of recollection suggests that their agency would be limited to cognitively fast simple tasks and would not extend to cognitively slow complex tasks. Admittedly, this is speculative. Research to date does not offer a definitive answer to the question of the degree of agency that those with SDAM retain. Yet it seems plausible that their capacity to initiate and execute action plans, as well as the type and range of these plans, would depend on the extent of damage to the hippocampal and cortical networks mediating this capacity.

Memory and Personal Identity

In the chapter entitled "Of Identity and Diversity" in *An Essay Concerning Human Understanding*, Locke asserts that the self is "a thinking intelligent being, that has reason and reflection, and can consider itself as itself, the same thinking thing, in different times and places" (Locke, 1690/1975, book II, ch. 27). For Locke, we are essentially conscious beings, and memory is a core property of consciousness. Being conscious in "different times and places" indicates that consciousness is necessary not only for personhood but also for personal identity, the idea of persisting through time as the same individual. Locke underscores the necessary role of what we now call episodic memory in his account of the criteria of personal identity: "As far as consciousness can be extended backwards to any past action or thought, so far reaches the identity of that person; it is the same self now as it was then; and it is by the same self with this present one that now reflects on it, that that action was done" (Book II, ch. 27). The act of reflection to which Locke refers can plausibly be equated with recollection in episodic memory. This is a necessary criterion of personal identity. It provides the critical link between our past and present selves.

Mark Rowlands describes this as the memory criterion of personal identity: "Necessarily, a person, P1, who exists at time T1, is identical with a person, P2, who exists at a later time, T2, if and only if P2 can remember an experience had by P1" (Rowlands, 2017, p. 93). More precisely, "an episodic memory had by person P1 of an experience had by person P2 is necessary and sufficient for the identity through time of P1 and P2" (p. 94). Identity is a transitive relation. If X = Y, and Y = Z, then X = Z. If my self at twenty is identical to my self at fifty, and my self at fifty is identical to my self at eighty, then my self at twenty is identical to my self at eighty. My capacity to recall experiences I had at different stages of my life is necessary for me to persist through time as the same person. Rowlands further claims that the relation of transitivity "obtains only between persons for which identity has been established in virtue of the one having episodic memories of the experiences of the other" (p. 97; Klein and Nichols, 2012).

These episodic memories are autobiographical memories. Yet if episodic memory is necessary for the transitivity of identity, and episodic memories fade over time, then there may be a limit to the temporal extent to which identity can hold. The storage and retrieval of episodic memories diminishes with the increasing temporal distance between an experience and a memory of it. This is a natural consequence of the brain and mind gradually eliminating older information to accommodate more recent information for adaptability to changing circumstances. This may have significant implications for identity in an extended life span.

The memory criterion of personal identity assumes that we are essentially psychological beings and that the relations that define identity are psychological relations. In contrast, some would argue that we are essentially biological beings and only nonessentially psychological (Olson, 1997, 2007; cf. Parfit, 2012). Psychological properties involving conscious and unconscious mental states emerge from a more basic integrated set of biological properties involving the body and brain. On this view, our conscious lives constitute only one phase of our more fundamental and more spatiotemporally extended biological lives. Persons, or human beings, are biological rather than psychological entities. This would mean that episodic memory has no necessary role in personal identity. Yet there is a subjective aspect to the experience of persisting through time. Episodic memory is a critical feature of the experience of mental time travel in recalling the past and projecting oneself into the future. Episodic memory is autonoetic and thus necessarily involves consciousness. Because episodic memory is necessary for personal identity, and this type of memory involves psychological properties, we need a psychological rather than a biological concept of identity to track the existence of a person through time.

Memory involves both neurobiological and psychological processes. Damage to the hippocampal complex and its connections to other limbic and cortical networks can result in memory impairment at the conscious level. It is largely the effects of brain function and dysfunction underlying episodic memory that determine whether personal identity is maintained, temporally disrupted or permanently lost. Neither an exclusively psychological nor an exclusively neurobiological concept of memory is sufficient to account for personal identity. Both concepts are necessary to account for how person P1 at T1 can be the same as person P2 at T2.

Derek Parfit's account of personal identity has been the most influential psychological account in the last forty years. According to Parfit, "to be a person, a being must be self-conscious, aware of its identity and continued existence over time" (Parfit, 1984, p. 202; 2012). He defines persistence in terms of what he calls Relation R with the right kind of cause, specifically the human brain (1984, pp. 211ff.). This relation involves psychological connectedness and continuity. In saying that identity as such does not matter, Parfit means that it does not involve anything over and above Relation R. Psychological connectedness consists in the holding of direct links between mental states such as beliefs and desires, between an intention and the action in which one executes it, and between an experience of an event and one's memory of it (pp. 211ff.). These connections can be stronger, holding over shorter periods, or weaker, holding over longer periods. Psychological continuity is the ancestral relation in psychological connectedness, consisting of overlapping chains of strong connectedness. These chains extend over longer periods than direct links between mental states. Unlike connectedness, continuity does not admit of degrees and is a transitive relation.

Strictly speaking, personal identity is based on the relation of continuity rather than connectedness. Parfit maintains that connectedness pertains to "what matters" (in our self-regarding and other-regarding interests) and is not equivalent to identity (pp. 245ff.). Intuitively, though, what matters to each of us and our conception of ourselves as individuals persisting through time cannot be separated. I accept the view of connectedness as a relation holding between mental states over shorter periods and continuity as a relation holding between mental states over longer periods. Both are matters of degree because the content of the memories in these relations changes over time. These mental states involve episodic memory of the past and imagination of the future from the perspective of a constantly changing present.

The role of memory in linking the past to the future from the present is critical for maintaining personal identity. Yet memories cannot be sustained indefinitely. This suggests that the temporal extent of personal identity, of how long one can exist as the same person, may be limited by memory. This can have important implications for our desires and intentions about the further future. It raises questions about the coherence of the desire for radical life-extension if a substantially longer life entails a substantial change in identity. Memory loss resulting in identity change can be an effect of amnesia or the natural temporal limit of the content of memory. In the case of radical life-extension, the person wanting an intervention in the body and brain for a much longer life may not be the same person who exists in the later stages of that life. I will return to this issue in the next section of this chapter.

Alternatively, one could adopt a narrative approach to identity to avoid the problem of weakening memory and psychological continuity in an extended life span. A person could form and maintain a life story by integrating the reconstructed past and imagined future into a unified self (Schechtman, 2014). The mental act of forming a narrative could overcome the effects of the passive experience of persisting through time. The story could be constantly updated as one lived further into the future. Here too, though, the unity and integrity of these mental states and thus the narrative itself would gradually weaken beyond a certain point. If episodic memory is an essential component of a personal narrative, and if the content of this memory weakens over time, then the narrative account of identity in an extended life span would be just as vulnerable to the limits of memory as the psychological continuity account. The fact that both models of identity have a first-person, autonoetic aspect suggests that they are not fundamentally different regarding the role of episodic memory in them. The capacity to recall experiences is necessary for both. Just as changes in the content of episodic memory over time weaken the connectedness and continuity of a person's conscious mental states, so too do these changes weaken the links in a personal narrative.

Some forms of amnesia can disrupt connections between one's present mental states and past experiences. Depending on the severity of memory loss, it can result in gaps in these connections and affect a person's experience of mental time travel. Brief episodes of amnesia followed by complete recovery of episodic and semantic memory do not cause a significant disruption of identity. Some patients undergoing electroconvulsive therapy (ECT) experience some degree of retrograde amnesia but gradually recover the memories. The induced seizures interfere temporarily with memory retrieval mechanisms in the brain (Goodman, 2011; Kroes, Tendolkar, Van Wingen et al., 2014). However, for those who experience permanent retrograde amnesia with loss of many autobiographical memories from ECT, there can be a substantial change in their identity. Temporary or permanent memory loss resulting in temporary or permanent disruption in the psychological relations constituting identity also occurs in some patients with epilepsy (Goldstein and Kapur, 2012).

Transient global amnesia (TGA) involves sudden onset of acute anterograde and short-term retrograde memory loss (Goldstein and Kapur, 2012, pp. 217ff.). TGA occurs in otherwise healthy people and involves the temporary loss of the ability to form new memories or recall more recent memories. It is also characterized by repeated questioning and repeated statements. "The episode will usually last for around five hours, with a gradual and complete recovery of anterograde and retrograde memory loss over the subsequent few days. The patient will invariably be left with an amnesic gap for the period of marked memory impairment" (p. 272). The duration of this gap will determine whether the patient retains or loses the degree of psychological continuity necessary to retain identity. In most cases, the unity of a person's mental states can accommodate a brief gap in continuity resulting from transient amnesia. TGA may not always have a clear neurobiological cause. Its onset may have a psychological component.

Some cases of TGA are associated with hypometabolism and decreased perfusion in frontal, temporal and parietal lobes. A significant number of people diagnosed with TGA based on these brain abnormalities have been subsequently diagnosed with primary progressive aphasia (PPA) (Graff-Radford and Josephs, 2012; Jun, Duffy and Josephs, 2013). Neurobiological and psychological abnormalities associated with TGA and PPA can impair different types of memory regulating the capacity to recall events, connect them with the present and communicate with others. These abnormalities can disrupt both identity and agency.

Transient psychogenic amnesia (TPA) has a psychological or psychiatric cause. It is "a direct response to major stressful circumstances" (Goldstein and Kapur, 2012, p. 272). Some forms of TPA result in fugue states, or what is now categorized as dissociative fugue (American Psychiatric Association, 2013, code 300.12). This is a dissociative amnesia "where individuals will lose their personal identity and wander for days and weeks away from their home" (Goldstein and Kapur, 2012, p. 273). Although the amnesia is reversible, the duration of the loss of episodic memory may cause a disruption in psychological continuity. In addition to loss of recognition of family and others, TPA may include loss of semantic and spatial memory about one's surroundings resulting in impaired orientation to space and time. There can be a significant disruption of identity and agency in these cases. In all forms of amnesia, the extent to which it disrupts personal identity depends on the duration and extent of episodic memory loss. The longer and more extensive the amnesia is, the greater will be the gaps in psychological continuity and the weakening of personal identity. In severe cases, the gaps will be wide enough to completely disrupt continuity and identity.

Severe permanent anterograde and retrograde episodic amnesia can undermine identity by disrupting the experience of mental time travel and the ability to construct a personal narrative of one's experience. Some amnesiacs like Kent Cochrane may retain an understanding of time and the experience of persisting through time. Still, it is not clear whether this capacity involves just semantic memory or episodic memory as well. Cochrane may have retained some temporal conscious capacities from facts and concepts without autobiographical memories.

A study of four amnesiacs with bilateral hippocampal damage showed that they could construct coherent personal narratives from impaired autobiographical episodic memory (Keven, Kurczek, Rosenbaum et al., 2018). This appears to refute research indicating that amnesia precludes the ability to construct these narratives. There are varying degrees of amnesia corresponding to varying degrees of dysfunction in brain regions mediating the encoding, storage and retrieval of episodic memory. It seems plausible that the four individuals in the study could construct personal narratives despite amnesia because their amnesia was not severe. Some of the hippocampal network and its projections to cortical networks were probably intact. He and Clive Wearing lost their capacity to form an autobiography because their amnesia was severe and resulted from extensive damage to hippocampal and cortical networks. Their inability to learn new information and retention of only a few emotionally significant episodic memories prevented them from combining their experiences into a unified and meaningful whole. One cannot construct a narrative consisting entirely of events experienced in the present tense. Although H.M. retained some semantic knowledge, he lacked the ability to understand how the facts he knew fit into a coherent diachronic pattern. H.M. was incapable of constructing a personal narrative from events he could not recall or an impersonal narrative from facts he could recall.

Reflecting on H.M., Corkin writes, "when we consider how much of the anxiety and pain of daily life stems from attending to our long-term memories and worrying about and planning for the future, we can appreciate why Henry lived much of his life with relatively little stress" (Corkin, 2013, p. 75). Yet a "world bounded by thirty seconds" (p. 75) is an extremely restrictive and disabling world by any measure. With his few episodic memories and fragmented semantic memories, "he had a general sense of identity. But, his deficient autobiographical knowledge – personal, unique episodes – meant that his self awareness was significantly limited" (p. 222).

Insofar as H.M.'s amnesia permanently disrupted psychological continuity, there is a sense in which he permanently lost his identity. The person who existed for many years following the surgery on his medial temporal lobes was distinct from the person who existed before the surgery. One can make this claim despite the seizures he experienced in the early stages of his life. The seizures significantly altered his neurological and mental states. But they did not alter these states to the same extent as the amnesia that affected him following the surgery. Corkin's comments point to an objective value judgment about his life. Given a choice between a life with some anxiety and pain from recalling the past and anticipating the future, and a life without anxiety and pain from a loss of episodic memory, most would choose the first over the second. The cost to agency and identity in a life without memory would be too great for it to have much value.

Permanent damage to the hippocampal complex from herpes viral encephalitis permanently disrupted the psychological relations necessary for Clive Wearing to retain his identity. Like Henry Molaison, Wearing's postinfection world was confined to the present. By losing the capacity to recall the past and project himself into the future, he lost the experience of persisting through time as the same individual. Because the infection destroyed the brain regions sustaining his episodic memory, Clive Wearing at T2 after the infection was distinct from Clive Wearing at T1 before the infection. His wife Deborah wrote in a 2005 memoir of the devastating effect that his anterograde and retrograde amnesia had on his experience: "It was as if every waking moment was the first waking moment. Clive was under the constant impression that he had just emerged from unconsciousness because he had no evidence in his own mind of ever being awake before ... 'I haven't heard anything, seen anything, touched anything, smelled anything,' he would say. 'It's like being dead'" (cited by Sacks, 2007, p. 110). In this case, losing identity through amnesia is effectively like ceasing to exist.

In normal brain function with intact hippocampal-cortical communication, episodic memory is a key feature of the psychological connectedness and continuity necessary to sustain personal identity over time. But the content of this memory can gradually weaken as it is replaced by the content of more recent memories of more recent experiences. The links between the content of one's present mental states and the content of one's past mental states weaken with increasing temporal distance between the conscious past and the conscious present. Beyond a certain point, these links dissolve. Consistent with constructive and reconstructive models of memory, the subject and her brain psychologically and neurobiologically constantly revise the information associated with the memory so that it is relevant to her present and future circumstances. Representations of more recent experiences are more relevant to these circumstances than representations of more remote experiences.

The decreasing adaptive value of memories of experiences in the distant past may explain the weakening of psychological connectedness and continuity over time. If only stronger connectedness and continuity can meet the criteria for the transitivity of identity, and if only memories of more recent experiences constitute these stronger relations, then personal identity can hold only for a limited period. The duration of memories is limited because of the need for the brain and mind to update information in promoting flexible thought and behavior. Information about what a person has experienced in the past is not intrinsically relevant but only instrumentally relevant to what a person experiences now and in the future.

As Schacter and Addis point out, information about the past is relevant because it enables us to imagine and plan for the future. For memory to be adaptive, where we are now and where we are going is more important than where we are coming from in our mental time travel. As part of this purpose, there is a gradual weakening of episodic memories from more detailed to more generalized representations of experiences. Eventually, older memories are replaced by more recent memories as stored information in the brain and mind. Retention of too many older memories could cause one to focus too much on the past and not enough on one's actual and foreseeable cognitive demands. Like the planks of Neurath's ship, though, goal-directed behavior requires a constant core set of memories that persists through these changes.

In hyperthymesia, strong connections between one's present awareness and autobiographical memories extending backward to childhood can ensure transitivity of identity. But these memories have limited value if they distract us from present and future cognitive tasks and interfere with planning and acting necessary to complete these tasks. The key issue is not how many memories we can retrieve, or how detailed they are. Rather, the key issue is that we have the capacity to reconstruct information about the past to enable effective reasoning and decision-making in the present and future. If one accepts a reconstructive rather than reproductive model of memory, and if one accepts the claim that reconstructing memories results in a weakening of the psychological relations necessary for identity, then identity may weaken and dissolve over time. The ability of *a* person to adapt to changing circumstances may be more important than persisting through time as the *same* person.

Memory, Identity and Interest in an Extended Life

There has been considerable debate about how the biological mechanisms of aging could be altered and allow us to have much longer lives (de Grey, 2004a, 2004b; Ocampo, Reddy, Martinez-Redondo et al., 2016; cf. Olshansky and Carnes, 2001; Olshansky, Hayflick and Carnes, 2002). In the future, it is possible that genetic reprogramming and manipulation of epigenetic factors could alter telomeres, apoptosis and other mechanisms in the aging process and substantially extend the lifespan. Theoretically, altering these mechanisms could avoid or delay the neurodegeneration in dementia and other intractable disorders affecting memory. This is speculative, given the current state of research in biogerontology. But let us assume that the human life span could be substantially extended, and that physical and mental functions could be intact for a prolonged period (Davis, 2018). The biological changes would have significant psychological implications for the role of memory in identity. Psychological connectedness and continuity involving episodic memories may hold to a sufficient degree over a normal lifespan of eighty-plus years. But they may not hold over a much longer life span (Park and Gutchess, 2005).

In a radically extended lifespan of two hundred years, P1 at T1 may be identical to P2 at T2, and P2 at T2 may be identical to P3 at T3. But if episodic memory and the psychological relations it constitutes weaken over time, then P1 at T1 may not be identical to P3 at T3. There would be successive but distinct selves. The psychological relations holding between them would involve very long temporal segments and be too weak to sustain identity. In that case, one could question a person's interest in having a much longer life and undertaking and completing projects within it. The person who realized the desire for a longer life and who undertook and completed these projects would not be the same person who initiated the biological process of life extension. P1 at T1 could not benefit from events at T3 if he was not identical to P3 at T3.

"Benefit" implies a comparison between two states of affairs in which the same person exists. She is better off in one of these states than she is in the other. Yet if the person who exists at T3 is not the same person who exists at T1, then it is unclear in what sense P1 at T1 would benefit from life extension. The reasons for a longer life would be self-defeating. Consistent with all accepted theories of episodic memory, if P3 lived for two hundred years, then she would have forgotten the experiences and interests of P1 at thirty years (Glannon, 2002a, 2002b; Harris, 2002). Because of the natural weakening and dissolution of memory and other mental states over extended periods, the person who had the much longer life would be distinct in all relevant respects from the person who wanted such a life.

When one has a desire to live into the further future, presumably this desire is not based on qualitative similarity between one's present and future mental states, including retrieval of memories of past experiences. It is based on numerical identity. The person existing in the future would be the same person who exists now. In wanting a radically extended life, it is not enough that the desires, beliefs, intentions and memories I had at an earlier time and currently have will be similar to the mental states of a person in the distant future. I do not assume that these mental states will constitute a related but distinct self. Instead, I assume that they will be *mine* and that it will be *me* who exists at the later time. I would want the episodic and semantic memories I retrieve now, the past experiences they represent and the future possibilities these memories enable me to imagine to be united by strong psychological connectedness and continuity.

Parfit claims that "connectedness matters more than continuity in theory and practice" (1984, p. 206). This includes self-regarding and other-regarding interests, and the realization of these interests. Over a very long period, connectedness weakens and fails to meet the transitivity condition necessary for personal identity in a strict sense. Both connectedness and continuity matter to us and our future interests. Some of the content of the original memory of my experience must be connected to and continuous with the content of my present mental states when I retrieve and update the memory. The original and updated content of the memory must also be connected to the mental states I anticipate having in the future. Insofar as episodic memory enables us to perceive the future as a mirror image of the past, how we acquire and process information about the past serves as a basis for how we acquire and process information about the future. Without some content of episodic memory remaining constant over time, there would be no neurobiological or psychological basis for these cognitive capacities. We may need only a limited amount of this content to adapt to the present and future. But it may not be enough to sustain identity. This suggests that identity and adaptability may work at cross purposes in a radically extended lifespan.

A certain degree of connectedness and continuity must hold between the content of a retrieved and reconsolidated memory and the content of the original encoded and consolidated memory to sustain identity and self-regarding interest in the distant future. This may not be necessary for adaptability to the environment. Because flexible thought and behavior require constant updating of memory to make it relevant to one's present and future circumstances, the change in content resulting from updating over an extended period may be so substantial that there may no longer be any connection between the representation in the original memory and one's present awareness.

In a normal lifespan, the content of distant memories and their connections to recent memories may weaken but not dissolve. These weakened relations may be enough to sustain personal identity over the entire life. In a radically extended lifespan, however, this content would substantially change, and so would identity. If the weakening and eventual loss of remote memory through updating serves an adaptive purpose, then changes in the content of memory over time may come at the cost of identity. This may preclude realization of an interest in a much longer life because, beyond a certain point, the self that had this interest at age thirty would no longer exist at age two hundred. There would be no rational grounds for the earlier self's interest in the distant future because a different self would exist at the later time.

Some might claim that biological manipulation of memory mechanisms would enable us to control the extent to which we could retain episodic memories. Their content would not weaken but remain constant over a much longer period. Yet if the adaptive purpose of memory requires optimal levels of remembering and forgetting, then memories that persisted for many years could disturb these levels by overloading the mind and brain with irrelevant information. There would be too much remembering and not enough forgetting to enable the person to learn new information necessary to anticipate the future and engage in future planning. Even if the duration of memories could be extended to maintain a person's identity, it could impair his agency. This is another sense in which choosing a much longer life could be self-defeating.

In the hypothetical scenario I have described, the identity of a structurally and functionally integrated human organism would likely be preserved. Its biological properties would be relevantly similar at age thirty and age two hundred. Only its nonessential psychological properties would change. But it is our psychology, particularly the role of memory in enabling us to project ourselves into the future, that grounds self-regarding interest in the future (Roache, 2018; Sauchelli, 2018). A long enough biological life would entail distinct psychological lives of distinct persons. If our interest in the future depends on a conception of ourselves as psychological rather than biological entities, and if our interest and the memories that sustain it extend only so far into the future, then there would be no reason for an earlier self to care about what might happen to a much later self. If one could not recall one's earlier interests, then any psychological connections between these two periods may be too weak to claim that the person who realized these interests was the same person who had them.

These considerations are pertinent to the debate on posthumanism (Gordijn and Chadwick, 2009; Agar, 2014). Hypothetically, brain implants and other devices could fundamentally change the properties that essentially define us. They could introduce a new ontological category of beings (Roache, 2018). This raises the question of the coherence of comparing the value of two states of affairs in which two qualitatively and numerically distinct kinds of being exist. If the purpose of converting humans into posthumans was to promote adaptability to a radically changing natural environment, then there would be no compelling reason to retain the psychological relations necessary for identity between present and future individuals. Or at least these relations would matter less in an extended lifespan than they would in a normal life span. In this scenario, the radical difference between present and future environments would probably mean that adaptability would depend more on memory of events in the recent past than on memory of events in the remote past.

We could call the scenario in which a person wanted to radically extend his lifespan "Methuselah's paradox," after the biblical character. Admittedly, the case of Methuselah is an exaggerated illustration of the role of episodic memory in retaining or losing personal identity over an extended period. Nevertheless, David Lewis employs it instructively to show how the weakening and dissolution of autobiographical episodic memory would result in psychological disconnectedness and discontinuity in a very long life:

Consider Methuselah. At the age of 100 he still remembers his childhood. But new memories crowd out the old. At the age of 150, he has hardly any memories that go back before his twentieth year. At the age of 200, he has hardly any memories that go back before his seventieth year; and so on. When he dies at the age of 969, he has hardly any memories beyond his 839th year. As he grows older, he grows wiser; his callow opinions and character at 90 have vanished almost without a trace by age 220, but his opinions and character at age 220 also have vanished without a trace by age 350. He soon learns that it is futile to set goals for himself too far ahead … For Methuselah, the fading out of personal identity looms large as a fact of life. It is incumbent on us to make it literally true that he will be a different person after one and a half centuries or so. (Lewis, 1976, p. 30)

This example suggests that self-interested concern about the future is based on the idea of persisting through time as the same person. The less psychological connectedness and continuity there is between my present self and a future self, the less rational it is for me to care about the future self, which would not be me. The extent to which the psychological relations constitutive of identity hold depends on how long episodic memories can be sustained. Because memories of more distant experiences are replaced by memories of more recent experiences in the brain and mind, there is a limit to the temporal extent of memories. For a person with a radically extended lifespan, there would be a weakening and gradual loss of psychological continuity between her remote past and her remote future. This would weaken her reasons for concern about what might happen to her many years from now. Parfit seems to support this point by saying that "connectedness is one of the two relations [the other being continuity] that give me reason to be specially concerned about my own future. It can be rational to care less, when one of the grounds for caring will hold to a lesser degree" (Parfit, 1984, p. 313). He further claims that "we can treat weakly connected parts of one [psychological] life, in some respects, or to some degree, like different lives" (p. 337).

Lewis's comment that Methuselah "grows wiser" as the connections between his earlier and later memories weaken suggests that this wisdom is not linked to identity. If this wisdom is a feature of adaptability to one's present and future circumstances, then it supports the idea that adaptability may come at the cost of identity in a radically extended life. This would be the case even in a life considerably shorter than that of Methuselah. A narrative approach to personal identity would not be immune to this effect. The connections between earlier and later segments of the narrative would weaken as more experiences were added in a longer life.

If one is involved in a very long-term project, then one's interest in it, and the project itself, may need revision to align with changing circumstances. This corresponds to the need to constantly update the content of memory so that it aligns with a person's actual milieu. Updating this content would not guarantee that one's earlier interests would be sufficiently related to one's later interests over the course of the project. On the contrary, changing the content of the memories would be more likely to weaken or dissolve this relation. This is turn could disrupt the unity and continuity of the project, as well as one's intention to sustain and complete it. These are possible effects of the limits of the human mind, and particularly episodic memory, in connecting the past to the present and future.

The role of the transcription factor CREB, protein synthesis and long-term potentiation in regulating the encoding, consolidation and storage of memory supports these claims. Normal gene expression through CREB regulates the integration of newer information and the elimination of older information in the brain associated with episodic and semantic memories. By limiting the amount of information at any given time, CREB ensures that the brain is not overloaded and not compromised in its functions. Twenty years ago, researchers distinguished between "activator" and "suppressor" genes regulating memory in *Aplysia*, *Drosophila* and mice in several studies (Abel, Martin, Bartsche et al., 1998; Lisman and Fallon, 1999; Tang, Shimizu, Dube et al., 1999). The first type of genes enables the encoding, consolidation and storage of long-term memory. The second type constrains the first to limit the amount of new memories in the brain. It can also remove existing long-term memories from circulation by interfering with retrieval mechanisms. This can prevent the reconsolidation necessary for a memory to remain stored. These optimal functions of memory-activating and memory-suppressing genes ensure a balance between learning and forgetting.

The results of these studies are still relevant today and can be applied to the role of memory mechanisms in the hypothetical case at hand. Anything less than a balance of memory activation and suppression can result in impairment or loss of the ability to process and apply information in performing cognitive and motor tasks. Given that there are limits to the amount of information the brain can hold at any given time, assimilation of new information requires the replacement or elimination of old information. This implies that long-term memories are stored for only a limited period and that the psychological continuity necessary for identity weakens over time. The transitivity of identity can extend only so far into the past and only so far into the future.

The ability to construct and reconstruct the content of episodic memory promotes flexible thought and behavior. Yet constantly updating this content gradually weakens the psychological relations between earlier and later selves. The adaptive purpose of memory may interfere with the continued existence of the same person over many years. If we are essentially biological beings, then substantial changes in memory would not affect our identities. There would be no substantial alteration of the biological properties that define each of us. These changes would indicate different phases of the existence of the same individual over time. But if we are essentially psychological beings, then substantial changes in memory in a radically extended life span would imply substantial changes in the psychological properties that define each of us. These changes would indicate the existence of different individuals over time. The difference in memory content between earlier and later times could disrupt identity and undermine reasons for self-regarding interest in the remote future. Still, what ultimately matters is not so much how memory affects identity but more so how it enables adaptability and survival in different environments. Failure to realize these goals may result in our demise, regardless of the type of beings we essentially are.

Dementia and Precedent Autonomy

In a normal lifespan, neurodegenerative diseases such as Alzheimer's resulting in dementia can cause a person to lose declarative and nondeclarative memory. An affected person gradually loses the capacity to encode, consolidate, store, retrieve and reconsolidate episodic and semantic memories. He also loses the capacity for spatial, working and prospective memory. His cognitive functions decline to the point where he can no longer reason and make decisions. Neurodegeneration permanently disrupts the connections between his present and past mental states and thus his identity. As dementia progresses, he also loses procedural memory and the ability to perform basic motor skills. Norman Cantor points out that "some people will confront Alzheimer's with a measure of resignation … For other people, like myself, protracted maintenance during progressive cognitive dysfunction and helplessness is an intolerably degrading prospect" (Cantor, 2018, p. 15; cf. Dresser, 2018; Sulmasy, 2018). They may take certain actions to avoid being in this state.

In the early stages of dementia, a person may retain enough insight and cognitive capacity to foresee that he will eventually lose this capacity and the ability to make decisions about medical care and other personal matters. He may judge that life with dementia is not worth living and form an advance directive to voluntarily decline life-sustaining nutrition and hydration when he reaches this stage. But he will not remember the wish he expressed in the directive and will not be the same person who expressed it.

How should we evaluate the wishes of a competent individual about actions that should be taken when he is no longer competent? Should others respect these wishes and act according to the directive despite the loss of memory and identity? In cases of moderate dementia, if a person's wishes conflict with wishes she expressed when she had full cognitive capacity, then which of these wishes would accurately reflect her interests and values? Should we give more weight to the interests and values of her later self than to those of her earlier self? Advanced dementia precludes the ability to remember an earlier directive about medical care in the last stage of life. But this does not mean that the directive ceases to have moral and legal force when one is demented. If earlier judgments about medical care reflect a person's interests and values about her life as a whole rather than different stages of her life, then these judgments could transfer to the later period when she has no capacity to recall or make decisions based on them.

These issues pertain to what Ronald Dworkin calls "precedent autonomy" (R. Dworkin, 1993, pp. 226–229). This concept refers to an autonomous advance directive from a competent patient regarding actions that would affect her when she is no longer competent. We should respect the expressed wishes of a person about how she should be treated if she became demented rather than wishes she might express in a demented state. The person's autonomy extends from the earlier time to the later time. Autonomous decisions and actions follow from motivational states that one endorses as one's own following a reflective process. As part of this process, a person may revise her motivational states or reject them and endorse others instead as the source of her actions (G. Dworkin, 1988; Mele, 1995). This requires a level of cognitive and emotional capacity that has been lost in advanced dementia.

Precedent autonomy presupposes people's capacity for what Dworkin calls "critical interests." These are "interests that ... make their life genuinely better to satisfy, interests they would be mistaken, and genuinely worse off, if they did not recognize" (Dworkin, 1993, p. 201). These interests reflect general beliefs, values and convictions about not just particular stages of one's life, but one's entire life. In this respect, the interests are time-neutral. Like autonomous intentions, decisions and actions, critical interests involve the cognitive capacity to critically reflect on and endorse or revise them. In contrast, "experiential interests" pertain to interests in having pleasurable experiences and avoiding painful or distressing ones. Experiential interests are time-sensitive in the sense that they involve a patient's present mental and bodily states and are indifferent to the past or future (p. 201).

Critical interests shaped by religious beliefs explain Jehovah's Witnesses' directives not to receive blood transfusions in medical emergencies when they are unable to make informed decisions. Their autonomy and right to refuse transfusions transfers from the time of the directive to the time when they are no longer competent. Similarly, the autonomous decision of a competent person who formulates a directive not to receive artificial hydration and nutrition if she becomes demented transfers from the time of the directive to the time when she is demented. The moral and legal force of the directive does not weaken with the loss of memory. The person is exercising her precedent autonomy because she has critical interests that obtain over all the stages of her life. Her directive reflects how she wants her life to go overall, not just when her cognitive capacities are intact.

Suppose that a person says that he would not want to continue living if he became demented. Living in such a state would be contrary to what he considers acceptable dignity and quality of life. He formalizes this wish in an advance directive. Once he becomes demented, however, he retains some communicative capacity and indicates that he wants to continue living. He cannot recall his earlier expressed wish.

Some may raise the question of whether the different attitudes reflect a genuine change of mind by the person, or whether the change was caused by the disease (McKerlie, 2013, pp. 186ff.). This may not be the appropriate question to ask, since presumably a person with advanced or even moderate dementia lacks the cognitive capacity to genuinely change his mind. At an advanced stage, the disease makes the person incapable of making or revising value judgments about his life. The cognitive impairment results in the loss of critical interests. It undermines his capacity to reflect on and endorse or revise considered wishes about the medical care he should or should not receive. He retains only an experiential interest in pleasurable experiences and avoiding pain and suffering. His earlier wish was autonomous because he was cognitively and emotionally capable of deliberatively forming, reflecting on and endorsing it. His later wish was not autonomous because the dementia caused him to lose this capacity.

In advanced dementia, the person is no longer an autonomous agent. His directive not to be treated if he becomes demented is an expression of his precedent autonomy, which retains its normative force over time. This gives more weight to the person's earlier wish than to his later wish. What he expresses in his advance directive reflects his critical interests in all stages of his life, not just any experiential interests he may have in the last stage of his life. If one accepts a psychological concept of personal identity, then in advanced dementia he loses his essential psychological properties and ceases to exist as the person who formulated the directive. Yet if his critical interests refer not just to his psychological life but also to his complete biological life, then these interests hold until he ceases to exist as a biological entity and is declared dead.

If caregivers acted on the competent person's earlier wish to withhold life-sustaining hydration and nutrition, then they would be respecting his precedent autonomy. There are legal questions about whether withholding or withdrawing hydration and nutrition from a demented patient would be permissible. Specifically, there are questions about what caregivers would be permitted or obligated to do in acting, or refusing to act, on standing requests for actions from people with advanced dementia. In principle, if a competent person expresses a wish not to be treated in a demented state, and if hydration and nutrition fall within the class of medical interventions that the person has a negative right to refuse, then he would have a right to refuse hydration and nutrition if he became demented. Precedent autonomy would justify this action. Caregivers would be obligated to respect the wish expressed by the person in his advance directive and his right to refuse these interventions.

This is controversial, though, because the person who expresses the wish not to have life-sustaining treatment is competent, while the patient from whom the treatment would be withheld or withdrawn is not competent. The person requesting the action would be different from the patient who would be affected by the action. Nevertheless, precedent autonomy holds despite these changes. It also preempts the question of whether the patient willfully refuses to eat, or whether the dementia causes him to forget to eat. It preempts the question of whether this second scenario could justify involuntary feeding. The advance directive would clearly express the person's reasons for forgoing nutrition beyond a certain point in his life.

The fact that the demented patient expressing a wish to continue living forgets the earlier contrary wish and is not identical to the cognitively intact person is not the crucial issue here. What is crucial is that the earlier wish reflects her critical interests. These interests persist through all the stages of her life and cannot be superseded by experiential interests that obtain in only the last stage. Forgetting her critical interests does not mean that they would no longer obtain beyond a certain point. The directive about how she should be treated later in life is sustained despite her inability to recall it.

Her critical interests "are still in force. They still apply to her and her life now when she is demented. She cannot consciously endorse them, but that is because she is unable to formulate those thoughts" (McKerlie, 2013, p. 186). Her critical interests refer to what she values or disvalues. If she disvalues living with dementia, then the directive to withhold or withdraw treatment should be upheld. "The patient did not abandon her previous values. Her view of her good was never retracted, and it continues to shape what is good for her now" (p. 196). Although episodic, semantic, working and prospective memory are necessary to formulate a directive about future medical care, the loss of these types of memory does not nullify her earlier advance decision-making. It is the anticipated and feared loss of memory and other cognitive functions in dementia that motivates the exercise of precedent autonomy in these cases.

One could argue that the demented patient's wish to continue receiving hydration and nutrition was an expression of her will to live. Still, it would not be clear whether this wish was consistent with a standing critical interest in living for as long as possible or an expression of an occurrent experiential interest in continuing to live. The default position would seem to be to act on the second interest and continue treating the patient. However, if the patient clearly expressed in an earlier directive that she would not want to be kept alive by any means if she became demented, then respect for her precedent autonomy would give more weight to withholding or withdrawing than continuing treatment. If the demented patient directly expressed a wish for continued life-sustaining treatment, then caregivers may be inclined to act on it. This may reflect a temporal and emotional bias in giving more weight to her present experiential interest against her lifetime critical interest. This bias, and the patient's noncompetent state, would not weaken the reasons against treatment.

More difficult to assess are cases where there is a conflict between the wishes expressed by a person when her cognitive capacities were intact and wishes expressed when she has early-stage dementia. At this stage, cognitive impairment may be mild to moderate, and the different wishes may be of the same person. When dementia has just started to develop, a person may retain enough cognitive capacity to reflect on her values and retain her critical interests. She may retain enough working and prospective memory and emotional capacity to make value-laden decisions. Dworkin acknowledges that the mildly demented patient may retain the capacity for autonomy and the right to exercise it (1993, pp. 226–229). If this patient decided to revise what she had expressed in an earlier directive, then this would be an autonomous override of her prior wish. She would be allowing her current autonomy to override her precedent autonomy. Accordingly, caregivers would be respecting her autonomy by acting in accord with the revised wish. Assuming that the patient had enough cognitive capacity to recall the earlier directive, this would mean genuinely changing her mind from not wanting life-sustaining care to wanting it.

Agnieszka Jaworska acknowledges that a patient with Alzheimer's disease may not be cognitively able to think in terms of her life as a whole. But the patient may have a critical interest and values that apply to the current temporal part of her life (Jaworska, 1999). The value she attaches to the last stage of her life may outweigh the value she attaches to previous stages. If we have asymmetrical attitudes toward the past and future and care more about what will happen to us than about what has happened to us, and if we are entering the last stage of life, then we likely will care more about this last stage than earlier stages.

An early-stage Alzheimer's patient may retain some working and prospective memory enabling some cognitive and emotional capacity for decision-making. If she expresses a deliberative and coherent wish to continue living despite having dementia, then this wish should be respected because it would reflect her critical interests. This wish could override any previous wishes about treatment expressed in an advance directive. Her revised critical interests would apply only to the last stage of her life. What matters to her from her present point of view is the remaining years of her life, not the years she already has lived, or how the quality of the remaining years may compare with the quality of the years in her life overall. For these reasons, we should give more weight to the wishes she expresses just after developing dementia and less weight to the wishes she expresses before developing dementia. It is a genuine change of mind and not a symptom of a disease from which she still is only mildly or moderately affected. These claims depend on a certain level of cognitive and emotional capacity for reflective thought and behavior, which would not likely remain beyond early-stage dementia. In advanced dementia, the patient loses this capacity and her autonomy. Earlier wishes about medical treatment expressed as precedent autonomy should override any wishes expressed in an advanced demented state.

These considerations support the claim that withholding or withdrawing life-sustaining nutrition and hydration may be morally justified in some cases of patients with advanced dementia. Some authors make the stronger claim that precedent autonomy may justify preemptive suicide in dementias such as AD (Davis, 2014). The thought of losing memory and other cognitive capacities is so distressing to some persons that they would prefer to end their lives before these capacities begin to deteriorate. Dena Davis notes that presymptomatic genetic and clinical testing could enable people to determine their risk of developing Alzheimer's. "Before the availability of these pre-symptomatic tests, even someone with a high risk of developing AD could not know if and when the disease was approaching. One could lose years of good life by committing suicide too soon, or risk waiting until it was too late, and dementia had already sapped one of the ability to form and carry out a plan. One can now put together what one knows about risk, with continuing surveillance via these clinical tests, and have a good strategy for planning one's suicide before one becomes demented" (p. 543).

The ability to form and carry out a plan implies that one would retain the necessary working and prospective memory to do this. One would lose this ability if one waited too long, forgot the plan and could not translate it into action. Waiting until one's dementia was advanced would reduce or eliminate one's agency. One would have lost an opportunity to preempt the loss of memory and the self by ending one's life before becoming demented. If a person knew that he was at high risk of developing AD and wanted to avoid having the disease, then preemptive suicide could be a rational choice. The cognitive capacity to consider this choice would also imply the capacity to change his mind and not follow through with the plan to end his life.

The idea of preemptive assisted suicide for AD or other types of dementia is more controversial than an advance directive expressing a wish to forgo life-sustaining care. There are significant moral differences between ceasing to provide life-sustaining care for dementia and assisted suicide to preempt it. A right to forgo or discontinue hydration and nutrition is a negative right, a right to refuse an intervention in one's body. A right to assistance from a physician to prescribe a lethal dose of barbiturates for preemptive suicide would be a positive right. It would be a claim by the patient on the physician to provide the means necessary for the patient to carry out his plan.

Positive rights are weaker than negative rights. Unlike the absolute obligation for others to respect negative rights, positive rights entail a weaker, imperfect obligation to respect them (Kant, 1785/1964; Thomson, 1990, chs. 2, 9, 11; Kamm, 2007, section II). This means that one need not always act on requests for assistance as expressions of positive rights. Specifically, this implies that medical professionals need not prescribe the drugs necessary for assisted suicide. Depending on the circumstances, they may be permitted but not required to do this. Any presumed obligation to the patient could be overridden by the physician's professional integrity and belief that assisting in suicide would not be consistent with best medical practices. In addition, standard justification for assisted suicide is that the patient experiences unbearable and interminable suffering. While a person who is at high risk of AD or has early symptoms of the disease may suffer from the thought of a bleak future, this is not the form or degree of suffering that is typically cited to justify assisted suicide.

Dementia may be a secondary symptom of neurodegenerative diseases. Some patients with Parkinson's disease develop Lewy body dementia in addition to motor symptoms (Savica, Grossardt, Bower et al., 2013). This could affect a patient's cognitive and emotional capacity to make a deliberative and persistent request for euthanasia or assisted suicide when the disease was advanced. Preemptive euthanasia or assisted suicide could avoid this situation. These actions could be justified in cases where dementia was likely to develop from a disease like Parkinson's. Yet the main reason for these interventions in this disease would not be to prevent dementia but to prevent or stop the suffering from the primary motor symptoms.

Advance euthanasia directives (AEDs) authorizing euthanasia and assisted suicide are permitted in the Netherlands. The case of a Dutch woman with Alzheimer's disease who was euthanized based on her AED raises questions about these practices on patients with dementia. A Dutch euthanasia review committee found that the physician in this case failed to follow due care requirements in causing the patient's death. As David Gibbes Miller, Rebecca Dresser and Scott Kim point out, the case is noteworthy because it is the first to result in a criminal investigation since the Dutch euthanasia law was enacted in 2002 (Miller, Dresser and Kim, 2019). They emphasize the need for oversight and enforcement of the criteria for euthanasia and assisted suicide in general and a critical assessment of the permissibility of AEDs for persons with dementia in particular (cf. Jongsma, Kars and van Delden, 2019). In principle, though, if a person is competent when she forms an AED to be acted on when she becomes demented, and if the AED falls within the scope of precedent autonomy, then euthanasia could be permissible for a patient with dementia.

If a competent person at an earlier time made an advance request for assisted suicide after she became demented, then presumably this would fall within the scope of her critical interests and precedent autonomy. But if the dementia undermined her physical and mental capacity to have a role in executing a previously planned action, then ending her life would be an example of euthanasia rather than assisted suicide. This would be even more difficult to justify if her dementia ruled out informed consent and a consistent deliberative request to die. It would be a case of non-voluntary euthanasia. In addition, the patient could not experience unbearable and interminable suffering once she was severely demented because she would not be aware of her condition. These considerations could preclude a right to die because of dementia and would prohibit a physician from performing euthanasia. Even if the patient had a critical interest in ending her life when she became demented, a physician would not be obligated or permitted to accede to a patient's request to die as an expression of this interest. There are limits to precedent autonomy and the extent to which a patient is entitled to realize her critical interests.

Rebecca Dresser objects to the idea of giving absolute priority to precedent autonomy over present interests for advance treatment refusals by patients with dementia. She says that "by the time the competent person's directive took effect, the patient would be unaware of the person's goals and values that drove her earlier choices" (Dresser, 2014, p. 551). But if the directive was an expression of the person's critical interests pertaining to her entire life, then the fact that she was unaware at the later time and had no memory of the directive would not weaken its moral and legal force. Still, Dresser claims that "a rule that ignored the demented patient's current interests would conflict with the moral (and legal) obligation to protect vulnerable people from harm" (p. 551), and that "such a rule would also conflict with the judgment that people with mental disabilities can have lives of meaning and worth" (p. 551)

If caregivers acted on a patient's directive to withhold life-sustaining care, then they would benefit the patient by respecting her critical interests. Although she would die as a result, they would not be harming a vulnerable patient. Her precedent autonomy would prevent her from becoming vulnerable by not allowing others to act contrary to what she wanted. Losing her memory and other cognitive capacities would cause her to forget her critical interests. But forgetting these interests would not mean that they did not apply to this last stage of her life. She would probably lack the level of awareness necessary for suffering. Analgesia and other palliative measures could prevent or alleviate any pain the patient might experience from withholding or withdrawing hydration and nutrition. This would respect the patient's dignity as well as her autonomy. While it is true that many people with mental impairment can have meaningful lives, there are others for whom a life with dementia would not have any meaning or worth.

The main issue is not an objective general judgment about people with mental disabilities but the unique experience of a person and how she perceives the prospect of memory loss and dementia. A request to withhold life-sustaining care would be an expression of the person's critical interests and precedent autonomy. Again, there are questions about the sort of obligation such a request would entail for others in executing the terms of the directive. But it may be in the best interests of the person considering her life as a whole. Dresser's objection to precedent autonomy may be convincing in cases of moderate dementia. But the patient would have to retain enough cognitive and emotional capacity, and enough working and prospective memory, to revise her values and decisions about treatment. The patient would have to retain some critical interests and not have just experiential interests. In that case, her precedent autonomy would transfer from the earlier time when she was competent to the later time when she was no longer competent.

Conclusion

Declarative and nondeclarative memory is necessary to form and execute action plans. Impaired episodic, semantic, working and prospective memory can interfere with different stages of decision-making and impair agency. Anterograde and retrograde amnesia can have this effect by limiting the amount of information consciously accessible to a person. Exceptional autobiographical and semantic memory can also interfere with agency by cluttering the brain and mind with unnecessary details. Impaired agency from too little or too much memory comes in degrees. Some individuals with episodic amnesia may retain some capacity to engage in some forms of decision-making. Severe amnesia can also disrupt the psychological connectedness and continuity necessary for personal identity. Loss of episodic and prospective memory may cause a person to lose the experience of persisting through time as the same individual. Here too, though, some amnesiacs may retain a sense of time and other temporal conscious capacities, albeit to a limited degree.

In normal memory function, there may be a limit to the extent to which one can retain identity. In adapting to present and future environments, the brain and mind replace information associated with memories of more remote events with information associated with memories of more recent events. Updating the content of memory is necessary to make it relevant to one's actual and anticipated circumstances in promoting flexible and adaptive thought and behavior. Yet the adaptive purpose of reconstructive memory may come at the expense of identity in a radically extended lifespan.

Competent persons may form an advance directive stating that they would not want life-sustaining care if they became demented. Living in such a state would not be consistent with their critical interests in how they want their life to go overall. Yet when they become demented, they have forgotten what they expressed in the directive when they were competent. Precedent autonomy makes the voluntary request transfer from the earlier to the later time. Their critical interests in all the stages of their life outweigh any experiential interests they may have in the last stage. Failing to recall an advance directive to withhold hydration and nutrition does not weaken its moral or legal force. People with advanced dementia eventually lose all memory functions and all aspects of their agency and identity. Because dementia is an irreversible disease, they enter a state of permanent oblivion before they die. They cease to exist as psychological beings before ceasing to exist as biological beings. Alzheimer's disease and other dementias are examples of the essential role of memory in our lives and how losing memory can undermine the meaning and value of the last stage of life.

Patients undergoing surgery with general anesthesia enter a state of temporary oblivion when they lose consciousness. They return from this state when the anesthetic effects diminish, and they gradually regain awareness. Some patients unexpectedly become aware during surgery despite being anesthetized. While some do not recall this experience, others form an explicit or implicit memory of it. This may result in neurological and psychological sequelae. Patients with dementia lose memories they would prefer to retain. Patients who recall being aware intraoperatively retain memories they would prefer to lose. The first group suffers from a disorder of memory capacity, the second from a disorder of memory content. I discuss empirical, epistemological and ethical issues surrounding anesthesia awareness with recall as a disorder of the second type in Chapter 3.

General anesthesia causes unconsciousness by suppressing neural mechanisms mediating arousal and awareness. Anesthetics also cause amnesia by disrupting mechanisms of memory consolidation. "Indeed, anesthetics are the most potent amnesic agents known, and understanding how anesthesia causes amnesia should provide clues to the basics of memory functioning. Thus, anesthesia can be a powerful tool for studying memory" (Hudetz and Pearce, 2009; Kerssens and Alkire, 2010, p. 47). The relations between anesthesia, consciousness and memory are complex. They are especially complex in cases where patients unexpectedly become aware intraoperatively and develop memories of this experience despite being anesthetized. Anesthesia awareness (AA) is the experience and explicit recall of sensory perceptions during surgery (Avidan and Mashour, 2013). Although these cases are relatively rare, they have been documented as among the most frightening experiences of pain, panic and complete loss of control of one's body (Cole-Adams, 2017). They often have serious long-term psychological consequences.

Techniques such as the bispectral (BIS) index measure anesthetic depth during surgery. But because anesthetic depth is not equivalent to the level of consciousness, these techniques cannot determine that a patient is or is not conscious. "Currently available indices are based on a probabilistic approach. Therefore, an index value provides a probability and is not 100% reliable. Thus, an index does not allow a clear separation of 'consciousness' from 'unconsciousness'" (Schneider, 2010, p. 128). In addition, the "occurrence of memory cannot be predicted during anesthesia, i.e., when information is being stored; it can only be detected after the end of anesthesia when it is too late to intervene" (p. 116). Evidence of explicit memory is based on the patient's report following surgery. Even more difficult to predict is whether a patient will develop implicit memories of events when she is unconscious. Because she could not consciously recall these events, memories of them could be inferred only indirectly from her behavior. Although certain drugs and techniques might modify these memories, in some cases they may be beyond the possibility of modification. Both explicit and implicit memories can harm people by their negative effects on their thoughts and actions.

After describing the effects on anesthetics on awareness and memory, in this chapter I consider whether and in what respects patients who experience AA can be harmed by it. I then consider psychological and pharmacological interventions that might prevent, weaken or erase memories of AA. Intervening to prevent memories of AA is problematic. When awareness is detected during surgery, memory consolidation may already be under way and difficult to reverse. Anesthetics or other drugs would have to be administered early in this process to prevent memories from forming. Intervening to erase memories already consolidated and stored is even more problematic. If a patient recalls and reports being aware days or weeks after surgery, then the memory of the experience may be resistant to modification. The longer the duration between awareness and a report of a memory of being aware, the more entrenched the memory is and the more difficult it is to alter or dislodge it from the brain and mind. For some patients, the result can be a treatment-resistant pathological memory.

Nevertheless, memory research and actual cases indicate that these memories may be prevented by administering consolidation-blocking drugs to induce anterograde amnesia. Crucially, this intervention would have to occur early in the process of encoding and consolidating the information to have this effect. Also, despite many obstacles, it is at least theoretically possible to erase memories of AA with reconsolidation-blocking drugs inducing retrograde amnesia.

The general ethical question about harm from AA with postoperative recall can be broken down into more specific ethical questions pertaining to preoperative, intraoperative and postoperative actions. Would the potential harm from AA with recall be significant enough to justify informing the patient of the possibility of these events preoperatively? Would it be appropriate for the anesthetist to ask the patient if he would want drug-induced amnesia in the event of unexpected awareness? How would any benefit to the patient from asking this question be weighed against the potential harm from the anxiety resulting from the thought of becoming aware during surgery? If awareness were detected intraoperatively, then would an anesthetist be justified in infusing a drug to prevent consolidation of a memory of the experience without the patient's prior consent? If a patient reported postoperatively that he was aware intraoperatively, then would the reasons for taking a memory-erasing drug outweigh reasons for retaining the memory? Should an anesthetist inform a patient of the possibility of implicit memories he may form while unconscious and of their possible effects on the patient's postsurgical behavior?

Consciousness, Memory and Anesthesia

Anesthesia is pharmacologically induced loss of consciousness (Kihlstrom and Cork, 2007; Langsjo, Alkire, Kaskinoro et al., 2012; Shushruth, 2013). This requires a brief explanation of the neurobiology of consciousness. Neurologists Fred Plum and Jerome Posner divide consciousness into two components: wakefulness, or arousal, and awareness of self and environment. The first component is mediated by the brainstem ascending reticular activating system (ARAS) and its projections to the thalamus. The second component is mediated by the ARAS, its projections to the thalamus and further projections to networks in the cerebral cortex (Posner, Saper, Schiff et al., 2007; Blumenfeld, 2009). Consciousness is a graded property (Alkire, Hudetz and Tononi, 2008, p. 880). The neural correlates of consciousness are neither fully on nor fully off but maintain a resting potential prior to their inhibitory or excitatory action (Tononi and Koch, 2008).

Whereas general anesthesia can quickly cause unconsciousness, emergence from unconsciousness to awareness is gradual and comes in degrees. This is because of the slow reengagement of thalamo-cortical and cortico-cortical networks mediating consciousness as the anesthetic effect diminishes (Mashour and Alkire, 2013). The gradual transition from unconsciousness to consciousness in anesthesia can make it difficult to know when a patient becomes aware following surgery and the level at which he is aware. This also makes it difficult to know whether a patient has a level of awareness enabling him to perceive and respond to pain during surgery.

Three main theories have been proposed to account for the neural correlates of consciousness. The global neuronal workspace theory says that consciousness arises from neural activity distributed across many brain regions (Baars, 1988, 1997; Dehaene and Naccache, 2001; Dehaene and Changeux, 2011). Neural synchronization theory says that consciousness arises from the synchronization of integrated neural activity as it changes across brain networks (Crick and Koch, 2003). The information integration theory says that consciousness is generated and sustained by connected networks of neural information in thalamo-cortical and cortico-cortical networks. As the degree of integrated information increases, consciousness emerges. As the degree of integrated information decreases, consciousness fades (Tononi and Koch, 2008; Koch and Mormann, 2010; Massimini and Tononi, 2018). This process depends on a balance between diversity and unity in the brain (Massimini and Tononi, 2018, p. 71). While these three theories overlap to some extent, the information integration theory may be the most intuitive account of consciousness. I use this account as the basis for my discussion of anesthesia and amnesia. Anesthesia causes unconsciousness by disrupting integrated information in the connections between the ARAS, thalamus and cerebral cortex. Depending on the dose, depth and duration, anesthetics do not always cause complete oblivion or do not cause it for as long as intended. This explains why some patients become aware during surgery.

Ned Block distinguishes between two types of consciousness: phenomenal consciousness and access consciousness. He defines the first as "experience" and says that "the phenomenally conscious aspect of a state is what it is like to be in that state" (Block, 1995, p. 227). In contrast, access consciousness consists in attention to information processing and its "availability for use in reasoning and rationally guiding speech and action" (p. 227; Block, 2007). While phenomenal consciousness consists in the mere fact of being aware, access consciousness consists in the ability to describe or explain this experience and use it as a basis for reasoning and decision-making. Presumably, whether a patient can process and use information about intraoperative awareness is less important than whether a patient can experience pain, anxiety or panic and develop an emotionally charged memory of this experience. There can be no access consciousness without phenomenal consciousness. The ability to report an experience presupposes having that experience. Indeed, Block claims that we sometimes experience pain without access consciousness (1995, p. 227).

Yet without a memory of being aware, there may be no way for the patient or surgical team to know whether he was aware. The only way of knowing this would be from a report from the patient. Access consciousness enabling this report would be necessary to confirm phenomenal consciousness in such a case. This requires a memory of the experience because the patient could not know of and report the experience unless he could recall it. This may involve short-term episodic memory during the surgery or long-term episodic memory after it. Memory is thus necessary for both subjective and objective confirmation of AA. But forming and retaining explicit and implicit memories of AA can harm the patient. Ideally, anesthesia prevents both awareness and a memory of being aware. In some cases, however, it cannot prevent either of these states.

What further complicates the relation between anesthesia awareness and recall is difficulty in distinguishing true from false memories (Mashour and Avidan, 2015, p. i25). They could be distinguished by assessing the patient's report of awareness against the actual events in the operating room. A patient's report of recollecting AA would not be the result of what he expected to occur during surgery because no one expects to become aware under general anesthesia. Nevertheless, anesthetists may not be able to clearly separate true and false memories in all cases of AA. While acknowledging this difficulty, I assume that, in most cases, patients' reports of being aware are based on accurate recall of actual experience

There are three goals of general anesthesia: loss of consciousness (or "hypnosis"), amnesia and immobility (Mashour, 2010). Sedation causes loss of some degree of consciousness, but not to the same degree as anesthesia. Although some researchers describe immobility as "muscle relaxation" (Kihlstrom and Cork, 2007), "drug-induced paralysis" may be a more accurate description. This involves a neuromuscular blocking drug in addition to the anesthetic, which causes unconsciousness but not paralysis. Other researchers add a fourth category of analgesia administered preoperatively with anesthesia (Langsjo, Alkire, Kaskinoro et al., 2012, p. 4936). Some patients become conscious during surgery despite general anesthesia. Awareness in these cases may involve being responsive without recall or recalling what one experienced while responsive. The estimated incidence of awareness with recall from prospective clinical trials is approximately between 1 and 2 per 1,000 cases (Avidan and Mashour, 2013; Pandit, Cook, Jonker et al., 2013; Errando and Aldecoa, 2014; Tasbihgou, Vogels and Absalom, 2018). This can be compared with the incidence of death from general anesthesia, which has gone from approximately 1 in 20,000 in the 1970s to approximately 1–2 in 200,000 in the twenty-first century.

The BIS and the end-tidal anesthetic-agent concentration (ETAC) monitor measure anesthetic depth from electroencephalography (EEG) recording of electrical signals on the scalp arising from the cortex. By recording signals in only cortical and not also subcortical regions, they fail to differentiate conscious from unconscious states in general anesthesia. They are not reliable indices of AA. Subcortical regions like the thalamus and limbic system become activated before cortical structures in the transition from unconsciousness to consciousness. A more primitive conscious state mediated by subcortical structures may be present in the absence of cortical activity (Langsjo, Alkire, Kaskinoro et al., 2012). Because cortical activation occurs gradually after subcortical activation, it may be unclear when a patient becomes aware. This underscores the limitations of EEG in measuring levels of awareness in patients with traumatic or anoxic brain injury resulting in postcoma disorders of consciousness (Owen, Coleman, Boly et al., 2006; Owen and Coleman, 2008). Although both EEG and fMRI can detect awareness in patients mistakenly thought to be vegetative, the fact that EEG records only cortical and not also subcortical signals makes it unable to identify patients at the first level of awareness.

Another way of measuring brain activity correlating with consciousness is with the perturbational complexity index (PCI). Adenauer Casali and coauthors explain that "PCI is calculated by (i) perturbing the cortex with transcranial magnetic stimulation (TMS) to engage distributed interactions in the brain (integration) and (ii) compressing the spatiotemporal pattern of these electrocortical responses to measure their algorithmic complexity (information)" (Casali, Gosseries, Rosanova et al., 2013). Yet if PCI only provides a more accurate measure of cortical activity and not also subcortical activity, then it is questionable whether it could confirm that a patient was aware or her level of awareness. Consciousness can only be inferred from electrophysiological measures of brain activity. These inferences are imperfect. Although increased metabolic activity in thalamo-cortical and cortico-cortical networks correlates with awareness, it is not identical to it and cannot establish that a patient was aware during surgery.

More advanced monitors may detect low-frequency electrical waves in the brain as a biomarker distinguishing consciousness from unconsciousness. The slow-wave activity saturation rate could be combined with the anesthesia saturation rate to make this distinction (Mhuircheartaigh, Warnaby, Rogers et al., 2013). Still, there would be variation in this rate among patients. Also, knowing this rate would not resolve uncertainty about the subcortical-mediated gradual transition from unconsciousness to consciousness. There may be cases in which this technique would not be able to detect or help to prevent conscious states involving less than full awareness. If none of the indices I have mentioned can clearly distinguish between conscious and unconscious states, then none can definitively show that a patient experiences fear or pain during surgery.

Anesthesia awareness is difficult to prevent and detect because of the incomplete knowledge of the effects of anesthesia on the neural networks generating and sustaining consciousness (Avidan, Mashour and Glick, 2009). Unintended awareness may be due to underdosing of an anesthetic or sedative or to falling concentrations of the drugs during or near the end of surgery. Depending on the concentration of the drug, some anesthetized patients can be awakened and follow commands through the isolated forearm technique without recalling the experience (Langsjo, Alkire, Kaskinoro et al., 2012). Being aware does not mean that a patient will recall being aware.

Inhaled or infused anesthetics cause amnesia by disrupting neural mechanisms regulating memory consolidation. They allow the encoding of memory but disrupt consolidation and thereby cause it to be forgotten (Veselis, 2006, 2017). This explains why some patients emerging from unconsciousness despite anesthesia intraoperatively later report having no memory of it. The amnesic effect may be stronger if the anesthetist administers a benzodiazepine such as midazolam preoperatively to reduce the risk of anxiety in case of awareness. The anxiolytic effects of benzodiazepines can also impair memory consolidation. These drugs can cause anterograde amnesia but not retrograde amnesia. They prevent memory formation but do not erase memories that already have been formed. Lower concentrations of anesthesia are effective in preventing explicit memory consolidation. Midazolam can cause anterograde amnesia in a dose-responsive manner when given preoperatively before general anesthesia (Bulach, Myles and Russnak, 2005). Administering a higher concentration of an anesthetic during surgery if awareness was suspected might cause a return to unconsciousness but would not induce or enhance any amnesic effects.

The effects of anesthetics on consciousness are different from their effects on memory. In the latter, they depend on connections between the hippocampal complex and neocortex, and between the hippocampal complex and amygdala (Phelps, 2004). Although anesthetics, sedatives and benzodiazepines disrupt consolidation of episodic memories in the hippocampus, they may not disrupt consolidation of fear memories in the amygdala and other limbic circuits. Subclinical doses of the common anesthetic propofol can cause anterograde amnesia by disrupting consolidation of some types of declarative memory. Yet one study has shown that, while propofol can prevent episodic memories, it "is relatively ineffective at suppressing amygdala activation at sedative doses ... These findings raise the possibility that amygdala-dependent fear systems may remain intact even when a patient has diminished episodic memory of non-fearful events. They may be of clinical importance in the perioperative development of fear-based psychopathologies, such as post-traumatic stress disorder" (Pryor, Root, Mehta et al., 2015, p. i104).

Even less emotionally charged vague memories of AA corresponding to a lower level of consciousness can result in generalized anxiety disorder. Fear memories that develop in the amygdala from AA may avoid the blocking effects of propofol on episodic memories in the hippocampal complex. It is possible that an anesthetic could disrupt memory consolidation in both the hippocampus and amygdala. But some patients have compromised conditions that can limit the type and dose of the anesthetic. In addition, research suggests that the hippocampus may have a role in consolidating not just explicit memories but also implicit memories (Kihlstrom and Cork, 2007; Veselis, 2015). Anesthetics infused preoperatively or intraoperatively might prevent the consolidation of episodic memories but not the consolidation of fear and implicit memories.

Some studies suggest that the brainstem rather than the amygdala mediates panic and fear responses to stimuli. In one study, three patients with bilateral amygdala damage experienced fear and panic after inhaling carbon dioxide (CO_2). The researchers concluded that interoceptive receptors projecting directly to the brainstem caused this response. Exteroceptive receptors projecting directly to the amygdala may not be the source of fear and panic in all cases. The brainstem controls these responses to stimuli internal to the body. The amygdala controls these responses to stimuli external to the body (Feinstein, Buzza, Hurlemann et al., 2013). Brainstem mechanisms reacting to physiological stress may cause fear and panic in patients who become aware. Like mechanisms in the amygdala, brainstem mechanisms may also be immune to the amnesic and anxiolytic effects of anesthetics and benzodiazepines. A patient could still experience fear and panic despite receiving these agents. Because brainstem mechanisms do not regulate explicit or implicit memory, this complicates the question of whether a patient experiencing AA forms and retains a memory of this experience.

Awareness, Memory and Harm

Pain and suffering from awareness and recall harm patients by defeating their interest in avoiding these experiences when they undergo surgery (Feinberg, 1986, pp. 31ff.). They also harm them by defeating their expectation of unconsciousness and amnesia. The harm associated with recall can be substantial because it can persist long after surgery. One study indicates that as many as 70 percent of patients receiving general anesthesia who experience intraoperative awareness and recall develop PTSD (Avidan, Jacobsohn Glick et al., 2011). This and other sequelae are more likely when patients are given a neuromuscular blocking agent. By paralyzing the patient, the drug makes her unable to communicate her experience to anesthetists and surgeons. Some who become aware may be traumatized by pain from being cut and cauterized. Waking up paralyzed and intubated can cause anxiety and panic in the presence or absence of pain (Tasbihgou, Vogels and Absalom, 2018). Conscious sedation entails similar risks. In a registry of patients reporting memories under sedation, 78 percent felt distress, and 40 percent had long-term psychological sequelae, including PTSD (Kent, Mashour, Metzger et al., 2013). These experiences may occur because of difficulty in titrating sedative drugs and delivering the right dose.

Suppose that a patient unexpectedly becomes aware during surgery despite being anesthetized. Although she cannot report her experience, a monitoring device detects brain activity suggesting awareness, and the patient is given a higher dose of the anesthetic. The patient quickly returns to unconsciousness and remains in this state until the surgery ends. She gradually regains consciousness. Postoperatively, she does not recall becoming aware during the procedure. Was she harmed by the experience?

How one responds to this question depends on several factors. If the patient was aware and felt pain, then she would have been harmed. Suffering in response to pain or from the anxiety or panic from becoming aware would also harm her. Failure to remember this experience does not mean that she was not adversely affected by it when it occurred. The patient could be harmed even by a brief period of awareness. Recalling it would cause additional harm by causing her to reexperience the pain and suffering in an emotionally charged memory of them.

Yet if analgesia prevented pain, and if benzodiazepines prevented anxiety and panic, then it is not clear in what sense she would have been harmed. Becoming aware during surgery without any adverse perceptions might not be distressing for every patient. A memory of an experience without pain, anxiety or panic would not necessarily be distressing either. To be sure, waking up without the ability to communicate one's experience could cause panic that anxiolytic drugs might not prevent. Even here, though, an anesthetist could use a priming technique by uttering certain words and making suggestions before and repeating them during surgery to promote a different unconscious response to stimuli if the patient became aware. Although some may question the efficacy of this technique, it could be a nonpharmacological way of preventing panic in patients who can tolerate only lower doses of anesthetics. Questions about harm must be informed by the fact that anesthetics can have different consolidation-blocking effects on hippocampus-mediated episodic memories and amygdala-mediated fear memories. Anesthetics and sedatives may block one type but not the other. In that case, an anesthetized patient may avoid an episodic memory but develop a fear memory of being aware during surgery.

Two distinct brain networks mediate pain perception: a sensory network consisting of the lateral thalamic nuclei and somatosensory and parietal cortices, and an affective network consisting of the medial thalamus, anterior cingulate and prefrontal cortices. These networks mediate physical, cognitive and emotional aspects of pain (Price, 2000; Rees and Edwards, 2009). Some responses to noxious stimuli seem to indicate activation of the sensory network. Athena Demertzi and Steven Laureys point out that most of the behaviors associated with noxious stimulation – eyes opening, quickened breathing, increased heart rate and blood pressure – are of subcortical origin and "do not necessarily reflect conscious perception of pain" (Demertzi and Laureys, 2012, p. 92). The inference from stimulation to pain may be more difficult to draw when a patient under general anesthesia receives a paralyzing drug that prevents any movement that would indicate pain perception. Still, one can infer that a patient who is conscious feels pain when he is being cut and cauterized. Demertzi and Laureys further divide pain into nociception and suffering. The first refers to the physical response to noxious peripheral stimulation. The second refers to the emotional response to the stimulation (p. 95). More precisely, suffering refers to a state of increased distress associated with events perceived as threats to the person (p. 95; Cassell, 2004).

Suffering in response to pain is more harmful than pain alone. A person may suffer without experiencing pain. An active nociceptive network and perception of pain does not imply that an affective network mediating an emotional response to pain is also active. Although pain and suffering are separable, the second usually follows the first. Yet if a patient felt only pain and had no emotional response to and no memory of it, then the harm would be limited to the period when she felt pain.

The relation between perceiving (feeling) and remembering pain raises questions about the acceptability of pain and judgments of harm associated with perceiving it. The demented and infants have impaired or underdeveloped capacity for long-term memory. Would their inability to recall a painful experience make exposing them to pain more acceptable than it would be to those who are able to recall it (Davidson, 2014)? Assuming that the intensity and duration of pain was the same, the inability of some people to remember a painful experience suggests that they would suffer less and be harmed to a lesser degree than those who could reexperience pain by recalling it. The demented and infants would suffer only when they were feeling pain. Nevertheless, when it is experienced, pain is equally harmful to all people, regardless of memory capacity. The lack of this capacity would not make exposing patients to pain any more acceptable than it would be for those who have this capacity.

In all cases, exposing patients to pain would be acceptable only if it was medically necessary and not of an unreasonable intensity or duration. But differences in memory capacity suggest that the amnesic effects of an anesthetic would be less significant for the demented or infants than they would be for people with the capacity for recall. Indeed, their impaired or absent memory would obviate the need for an amnesic drug if they had to undergo major surgery. Because they could not recall pain, anesthesia and analgesia without an amnesic drug would be sufficient to prevent them from being harmed.

As part of the emotional response to pain, suffering may also be caused by anxiety in anticipating pain (Ploghaus, Tracey, Gati et al., 1999).

Anticipation of pain can cause mood changes and behavioral adaptations associated with suffering. To suffer in this way, a patient who became aware during surgery would have to form and retain a memory of feeling pain when he was aware. Even a very short-term memory of a painful experience can cause fearful anticipation of a continuation or new experience of pain. Suffering could be avoided by preventing the capacity to recall the pain and thus the ability to suffer in response to it. Without a memory of the experience, one could not anticipate additional pain. This could occur through the amnesic effects of a preoperatively infused anesthetic or a rapid-action memory consolidation–blocking drug infused intraoperatively when the patient became aware.

The cases with the greatest potential for harm are those with full intraoperative awareness and complete recall. These typically involve patients who are physiologically compromised and can tolerate only lower doses of anesthetics. Decreasing the risk of anesthetic morbidity increases the risk of intraoperative awareness with postoperative recall. The risk of harm is greatest with neuromuscular blockade. Immobility during surgery avoids interference with intubation and the ventilator that breathes for patients under general anesthesia. Yet awareness can be especially traumatic because the immobilized patient is unable to indicate that she is aware.

If an intubated and paralyzed patient becomes aware, then he may panic from the feeling of suffocation in not being able to breathe on his own. This feeling and the experienced loss of bodily control can be intensified by the temporary loss of proprioceptive and somatosensory feedback from the body to the brain, given the paralyzing effect of the drug. Local anesthetics blocking pain pathways in limbs have more limited effects on these types of feedback because they involve only parts of rather than the entire body. The risk of panic from AA could be minimized if the anesthetist preoperatively informed the patient that he would be given a paralyzing drug. This would depend on the patient forming and retaining a memory of this information, which he could access if he became aware. It assumes that the anesthetic, combined with an analgesic and anxiolytic, would leave the patient's capacity for episodic memory consolidation and retrieval intact. This explicit information would be different from the implicit information in suggestive words involved in priming. It is possible that both types of information could help minimize the risk or reduce the degree of harm.

Still, awakening without being able to report it would be an emotionally charged experience for any patient. This would be an example of the adverse psychological effects from having phenomenal consciousness without access consciousness. The experience can trigger the release of high levels of norepinephrine, which could combine with other mechanisms of consolidation and result in an intractable traumatic memory of it. This could firmly embed the information associated with the memory as it transitions from encoding in the hippocampus to consolidation and storage in the amygdala (McGaugh, 2004; Parsons and Ressler, 2013; Sillivan, Vaissiere and Miller, 2015). Fear memories in this brain region may be immune to the memory-blocking effects of anesthetics infused preoperatively and have significant psychological sequelae affecting the patient long after surgery. This is what occurs in the estimated 70 percent of patients experiencing anesthesia awareness with recall who develop PTSD.

Allowing different periods and levels of awareness during some procedures involving general anesthesia or sedation may avoid or mitigate unwanted intraoperative and postoperative outcomes. This would allow interaction between the anesthetist and patient and do more to prevent adverse neurological and psychological effects. Informing the patient preoperatively that she will be awake during part of the procedure could generate an expectation of awareness that might allay any anxiety or distress she might experience if she becomes aware. In expected awareness, fear and panic systems in the amygdala and brainstem might be primed for the experience and not activated to the same degree as they would be in cases of unexpected awareness. This would depend on the patient's preoperative level of anxiety. The information could trigger or exacerbate a negative response from a patient with an anxiety disorder. Besides, allowing patients to be awakened would not be feasible when they were being cut and cauterized. So, the instances in which controlled awareness during surgery occurred would be limited.

In outpatient procedures such as biopsies and endoscopies involving amnesic sedatives, a physician may tell a patient that she may feel minor pain or discomfort but not have a memory of it. This may influence the patient's expectation, allay any anticipatory anxiety and mitigate an emotional response to any pain she might feel during the procedure. It would not ignore the harm from feeling pain but could limit any subsequent adverse effects. These issues are significantly different in surgery with general anesthesia. It would not be appropriate to tell a patient that any pain felt during surgery would not harm her because she would not remember it. The anesthetist could not sincerely tell this to the patient given the possibility of recall of AA. Pain from a major surgical procedure would be greater than it would be in these other procedures, as would the patient's negative emotional response to the pain and memory of it.

Given the possibility of unexpected intraoperative awareness, and recall of being aware, should the anesthetist inform the patient of these possibilities preoperatively? Some would argue that the high probability of harm from the anxiety in response to the information, combined with the low probability of awareness and recall, would provide a reason against the anesthetist telling the patient of events that might occur. There may be reasons for doing this if the patient is deemed at high risk of awareness because of her condition. Even here, though, the value of the information would have to be weighed against the disvalue of the anxiety it could generate and the unpredictability of awakening during surgery. Potentially harmful effects from the information at the conscious level might be cancelled by potentially beneficial effects of suggestion at the unconscious level. Still, priming may have either positive or negative psychological effects in the patient. Because it operates at an unconscious rather than conscious level, there is no way of knowing what its effects might be in every case. Others would argue that patients should be informed of the probability of any untoward events that might occur during surgery. They might argue that, while the probability of AA may be low, the magnitude of harm from having this experience and living with a traumatic memory of it would be high and accordingly should be noted.

An obligation to disclose this information in principle would presuppose that there would be intraoperative interventions that could prevent awareness and psychological sequelae from recalling it. Without a pharmacological intervention that could quickly return the patient to unconsciousness, as well as prevent a memory of being conscious, being told of possible events without any means of preventing or reversing them would not be in the patient's best interests. In that case, the reasons for informing a patient of the possibility or probability of AA would be weak. But if these interventions were available, then a patient at high risk of AA could conditionally consent to them based on this information.

Whether a patient should or should not be informed of the probability of intraoperative awareness with recall may depend on how one interprets informed consent and respect for patient autonomy. A weaker interpretation of consent and autonomy would warrant informing the patient of only events likely to occur during surgery. This may rule out any mention of intraoperative awareness or postoperative recall. Not all patients want to be fully informed of all events that might occur during a procedure. Some may want more information than others. The amount of information about AA and recall that should be disclosed would be based on discussion between the anesthetist and the patient before surgery. Justification for discussing the possibility of awareness would depend on whether the patient was at risk of becoming aware. Reasons for discussing the possibility of recall would require information about the amnesic effects of anesthetics and how, in some cases, memories of what occurs in the operating room may form despite anesthesia.

In cases of suspected awareness, it can be difficult to predict whether a patient will form a memory of her experience. Memories emerge from levels of information integration in the brain but are not identical to the neurophysiology associated with this information. While emotionally arousing memories may correlate with an increase in neural activity detectable on functional neuroimaging, this activity alone will not reveal the content of these memories. There is variability among patients in how their brains encode, consolidate and store information about their experience. Some but not others may encode and consolidate memories of events occurring during surgery. There is also variability among patients in whether they retrieve stored memories, as well as when they retrieve them. These differences may be functions of transcription factors in the brain like CREB, as well as LTP and the effects of stress hormones like cortisol and norepinephrine. As noted, some patients postoperatively deny being aware during a procedure despite following commands and carrying on a conversation with the anesthetist and surgeon (Veselis, 2006; Langsjo, Alkire, Kaskinoro et al., 2012). Cases of intraoperative awareness without recall show that retrospective oblivion is not proof of unconsciousness.

Absence of evidence does not constitute evidence of absence. But absence of evidence does not constitute evidence of presence either. Lack of objective signs of awareness does not prove that the patient is or was aware. Yet without any evidence, there are no grounds for believing that a patient experienced AA. Neurophysiological criteria necessary to make a conclusive statement about consciousness or unconsciousness might not be available. Even if evidence of these states were available, it would be at best imperfect and incomplete. The only direct way of knowing whether an anesthetized patient was aware would be from her first-person postoperative report of awareness based on her memory of the experience. This assumes that the patient's memory was not imagined or false but an accurate representation of what occurred during surgery. The assumption itself may be questionable given the fallibility of memory as a representation of past events. It underscores the earlier point about the importance of access consciousness in being able to communicate experience. Specifically, it underscores how this ability depends on memory, and how phenomenal consciousness without access consciousness can preclude interventions to prevent or mitigate psychological sequelae from AA with recall.

There are three main questions here: empirical, epistemological and ethical. First, did the anesthetized patient become aware during surgery, and did she form a memory of it? Second, how could the anesthetist or surgeon know that the patient was aware intraoperatively, and how could the patient know whether she was aware if she had no postoperative memory of it? Third, how would the patient be harmed by awareness and harmed further by a memory of being aware? The respects in which she would be harmed are obvious. Pain and suffering from being cut and cauterized, panic from awakening while paralyzed and intubated and recalling these experiences would all harm the patient by allowing her to have experiences she has a clear interest in avoiding. An answer to the ethical question depends on answers to the empirical and epistemological questions. Judgments about being aware, forming a memory of awareness and how one is harmed by both depend on knowledge that the patient was in fact aware and formed a memory of it. There may be indirect evidence for these neurological and psychological states, but not the direct evidence necessary to provide straightforward and conclusive answers to these questions (Glannon, 2014a, 2014b).

The Ethics of Induced Amnesia

Anesthetics, sedatives and other compounds given preoperatively can prevent memory consolidation and thus cause anterograde amnesia of events occurring during surgery. Yet some patients develop memories of these events despite these drugs. This may be due to different effects of the drugs on memory mechanisms in the hippocampus and amygdala, among other factors. One question is whether an intraoperative intervention could also prevent memory consolidation and cause anterograde amnesia. A more speculative question is whether a postoperative intervention could erase these memories and cause retrograde amnesia. Propofol and benzodiazepines can create a virtual lesion in brain nuclei associated with the memory trace by disrupting the information constituting the trace (Veselis, 2015, p. i15). Local anesthetics create temporary lesions in efferent, afferent, somatosensory and proprioceptive pathways. General anesthetics with amnesic effects could create permanent lesions in nuclei of circuits mediating episodic memory. The lesions would be permanent in the sense that the drug would prevent the reintegration of information necessary for the formation and retention of the trace.

Whether memory formation could be disrupted when a patient became aware would depend on the stage of consolidation at the time of the intervention. Whether a memory of this experience could be erased days or weeks afterward would depend on the ability of the drug to disrupt reconsolidation when it was retrieved. I will consider drugs that might disrupt consolidation intraoperatively and then drugs that might disrupt reconsolidation postoperatively. If an anesthetized patient showed signs of awareness, then a higher dose of the anesthetic could be given to again cause unconsciousness. Yet too much anesthesia or sedation can be as harmful as too little. In addition to potentially adverse effects on cardior-espiratory function, a higher concentration of anesthesia could interfere with cognitive and emotional functions and result in mental impairment that could last for days, months or even years after the surgery.

Deep sedation, for example, has been associated with a high incidence of postoperative delirium. The incidence is highest among the elderly. For some patients, this neurological condition may develop into dementia (Pandharipande, Girard, Jackson et al., 2013). Anesthetics such as ketamine can produce intense technicolor dreams. These may be more frightening than pleasant (Davidson, 2014, p. 560). Leaving these effects aside, preoperative infusion of midazolam, ketamine, propofol or other drugs could prevent the consolidation of a memory of awareness when they do not completely suppress consciousness. Significantly, while anesthetics can interfere with memory consolidation and cause anterograde amnesia, they do not disrupt memories that have been consolidated and reconsolidated and therefore do not cause retrograde amnesia. A higher concentration of an anesthetic would not have this effect.

One possible exception to this point is xenon gas, which is an anesthetic shown to disrupt reconsolidation of fear memories in rats (Meloni, Gillis, Manoukian et al., 2014). It would be premature to claim that inhaling this gas would solve the problem of anesthesia awareness with recall because this effect has not yet been tested in humans. If awareness is detected intraoperatively, then a drug would have to disrupt consolidation when it was already in progress to prevent memory formation. In many cases, consolidation may be too far advanced for any drug to stop or reverse it. In some cases, however, pharmacologically induced anterograde amnesia may be possible.

Consider the hypothetical case of a patient with a compromised medical condition undergoing major surgery. A lower concentration of anesthesia is given to minimize cardiovascular and respiratory load. If this entailed a risk of AA, then the anesthetist could ask the patient before the surgery if he would want pharmacologically induced amnesia in case he became aware (Tenenbaum and Reese, 2007; Shepherd, 2014). This assumes that there would be a drug that could disrupt consolidation of a memory of the experience. The anesthetist and patient would have to weigh the potential harm from preoperative anxiety about the prospect of awareness against the potential long-term harm from a traumatic memory of the experience. They would also have to weigh these considerations against potential adverse effects of the drug.

Disclosing this information to the patient could give him some control over the memory process by allowing him to make decisions that could influence the sequence of events within it. The competent patient's autonomy and partial control over this sequence would transfer from the time he consented to or refused a proposed intervention to the time when he was aware but immobile and unable to make decisions. Although he would lack access consciousness at the later time when he was phenomenally conscious, he would have access consciousness at the earlier time when communicating his wishes to the anesthetist. The moral and legal force of this communication would extend from his preoperative to his intraoperative state.

Unlike advance directives for medical care in dementia, access consciousness in this case would be an example of precedent autonomy holding for a shorter period. But the moral and legal force of advanced consent would be similar in both types of case. If becoming conscious was unforeseeable because the patient was not deemed at risk, then administering the drug intraoperatively would be justified to prevent psychological sequelae from remembering it. Provided that the drug did not interfere with the functions of other memory systems and did not have any adverse effects, most patients would not object to the intervention. Some might judge a failure to intervene in such a case to be a negligent omission. But this would require clear evidence that the patient was at high risk of awareness and recall, in which case it would not be unforeseeable.

The question of harm arises not only from the pain, anxiety or panic from being aware but also from events occurring around the patient. These events can exacerbate the harm from these experiences. The emotionally charged content of the memory may include words or actions by the surgeon or anesthetist. Some patients may hear disturbing comments from the surgeon or anesthetist that were not intended to be heard. Suppose that a patient with colorectal cancer undergoes a resection of the large intestine. The surgeon comments on the poor prognosis after closing the incision. Unbeknownst to the surgeon, the patient hears the comment as the anesthetic wears off and he gradually regains full consciousness. A memory of these words could add to the suffering the patient has endured with the disease and will continue to endure after the surgery. If the anesthetist had reason to believe that the patient was aware and heard the comment, then infusion of propofol or midazolam to disrupt consolidation of an explicit memory of the comment might prevent postoperative harm from the memory. This action would have a higher probability of success if the patient had been aware for only a very brief period, and the drug was infused at the initial stage of consolidation with a rapid disruptive effect. The patient would likely form and retain the memory if the drug was infused at a later stage.

An anesthetist performed a similar action in an actual case. A woman was undergoing a biopsy performed by an orthopedic surgeon for a suspicious-looking lump near the end of her collarbone. Like most biopsies, she received only local anesthesia and was fully conscious. Still in the surgical suite after the tissue had been sent to pathology, she heard a comment from a pathologist over the intercom that the tissue was cancerous. The comment provoked an emotionally charged outburst from the patient. The attending anesthetist immediately infused the patient with propofol intravenously in her arm. He did this not to cause deeper sedation or re-sedate her, but to prevent consolidation of a memory of the comment and its disturbing content. The patient could not recall the experience after the procedure (Haig, 2007; Kolber, 2008). The propofol was effective in blocking consolidation because this process was in its initial stage and the memory trace was not yet in a stable state.

As an action intended to prevent a memory of an unforeseeable traumatic event with potentially harmful consequences, the propofol infusion seems justified in this case. One might object that the action was intended to correct the pathologist's apparent negligence in commenting on the tissue sample. It would be unjustified because it would be an attempt to cover up a wrongful act. But the comment was inadvertent and made by someone other than the physicians performing the biopsy and infusing the drug. It would not be objectionable for one agent to perform an action to prevent a harmful outcome that would have resulted from an action performed by a different agent. Although the patient was conscious when she heard the pathologist's comment, her heightened emotional state in reaction to it might have impaired her cognitive capacity to make a deliberative informed decision about whether to have the infusion. Also, given the narrow time interval for disrupting memory consolidation with propofol, asking the patient if she wanted this intervention would have allowed the consolidation process to develop and thwarted the intended effect of the drug. For these reasons, informed consent in this situation would not have been feasible. Because both the pathologist's comment and the patient's emotional reaction could not be anticipated, preoperative discussion of a possible propofol infusion would not have been feasible either.

Whether the physicians in this case were obligated to inform the patient of the action after the fact, or how the information should have been presented, are separate questions. They did not inform her. She died some time after the procedure from the cancer. The physicians later expressed some doubts about whether not informing the patient of the propofol infusion was the right thing to do. It is questionable whether this information would have benefited the patient by enabling her to make more effective medical decisions about her cancer treatment or nonmedical decisions about her life. Nevertheless, as an expression of her autonomy, a patient would want to be informed of any intervention in her body. Failure to do this on grounds of therapeutic privilege would be unduly paternalistic and violate her autonomy.

This case raises the question of how much medical information a patient needs after a procedure to make decisions that are in her best interests. Some patients want more information than others for this purpose. The default position would be to inform the patient, though this would have to be qualified by contextual factors unique to the patient such as her emotional state. If there were an obligation to inform the patient of the propofol infusion after the procedure, then would this include an obligation to explain that this drug may have different effects on memory mechanisms in the hippocampus and amygdala. Would this obligate the physician to explain that the drug might not disrupt consolidation of a fear memory of her experience and that she might recall it despite the infusion? Given the patient's circumstances, this information could fall beyond the scope of the physician's obligation to disclose it as medically necessary and not in her best interests. The physician could invoke therapeutic privilege and not provide this information. It would likely be more harmful than beneficial for a patient having to deal with the emotional aspects of terminal cancer.

Cases in which a drug was administered or taken postoperatively to induce retrograde amnesia of an intraoperative experience may seem less ethically problematic. The patient would have a choice about whether to accept or refuse the drug in exercising her autonomy. There are many empirical questions about the possibility or erasing memories. I explain the neuroscience and cognitive psychology behind memory erasure in more detail in Chapter 4. To erase an episodic or fear memory, the intervention would have to take place when the memory was being retrieved before reconsolidation. Memories that have been consolidated and stored need to be retrieved and reconsolidated to remain in storage. Retrieval makes memories labile and susceptible to modification. This puts them in a destabilized state allowing for updating before they are restabilized, reconsolidated and re-stored. Because reconsolidation requires protein synthesis, infusion of a protein synthesis inhibitor such as anisomycin into the amygdala could block this process. It could remove the memory trace from storage in the brain when a person was retrieving it.

Although research into using these drugs for reconsolidation blockade is at an early stage, they may have limited retrograde amnesic effects. Depending on the dose, they could also be toxic. In addition, the longer the period between the initial encoding and consolidation of a memory and its retrieval, the more firmly embedded the memory is in the brain and the more difficult it is to disrupt reconsolidation and dislodge it (Parsons and Ressler, 2013).

In fear memories, high circulating levels of cortisol and norepinephrine can activate excitatory synapses and enhance storage of a representation of a fearful experience through consolidation and reconsolidation in the amygdala (Krugers, Zhou, Joels et al., 2011). This can make the memory resistant to modification. A patient who does not recall being aware until weeks after the incident may not be able to erase the memory by any means. This is significant because only 50 percent of affected patients report anesthesia awareness immediately after surgery. Many do not report it until a month after the event, or even later (Goddard and Smith, 2013). Drug-induced retrograde amnesia for a memory of AA would be more difficult to achieve than drug-induced anterograde amnesia by preventing memory consolidation preoperatively or disrupting it intraoperatively.

Let us assume that a protein synthesis inhibitor taken shortly after a patient reported being aware would have a retrograde amnesic effect and safely erase a memory of the experience. The limited time in which the drug could be effective could constrain the patient's deliberation and decision-making. The patient would have to process complex information about the potential benefits and risks of the drug within a relatively brief period. As an experimental drug, information on its effects on both declarative and nondeclarative memory systems would be incomplete. Concerned about the psychological consequences of a long-term traumatic memory and the brief period for reconsolidation blockade, the patient may make a hasty decision without duly considering all the possible effects of the drug. It is also possible that a patient would not recall being aware until weeks after the episode. In that case, he could have more time to deliberate whether to take the drug. Yet by then drug-induced retrograde amnesia may no longer be possible. The neurobiological challenges with postoperative memory modification indicate a clear preference for preventing traumatic memories of AA from forming in the first place. This could come from anesthetics with more powerful amnesic effects administered preoperatively or intraoperatively.

There are many questions about the safety and efficacy of memory-modifying drugs. These are still at an experimental stage. Those aimed at blocking consolidation would likely be more effective than drugs aimed at blocking reconsolidation. But a patient who did not recall and report an episode of awareness until some time after the surgery would still have a strong desire to prevent harm from the memory and a strong reason for taking a drug to erase it.

I have noted how comments and other actions by physicians and other medical professionals in the operating room can harm patients. In an anonymous essay published in 2015 in the *Annals of Internal Medicine*, the author describes two disturbing cases (Anonymous, 2015). In one case, a male doctor made a sexually explicit comment while preparing an anesthetized woman for vaginal surgery. In a second case, a physician treating a Hispanic woman for hemorrhage after giving birth danced and sang a Mexican folk song. Actions like these violate patients' dignity and right to be respected as ends in themselves and not merely as means, as an object of a medical procedure (Kant, 1785/1964). They can be harmed by these actions even if they are not aware of them. They retain their dignity and rights when they are unconscious (Feinberg, 1986, pp. 65ff.). These verbal and physical actions are especially reprehensible because the patients are in a vulnerable situation and physicians are abusing their professional power. The administration of amnesic drugs might incline some physicians to perform actions in the presence of unconscious patients that they would not perform in the presence of awake patients. The problem would not be with the drug if it was medically indicated for the procedure. Rather, the problem would be with the unprofessional and unethical behavior of the physicians. There would be no grounds for avoiding the drug because of this behavior, which should be assessed separately from the effects of the drug on patients.

Having an explicit memory of intraoperative awareness, pain or panic can cause additional harm to these patients by causing them to reexperience them. If these patients could take a protein synthesis inhibitor or other drug to erase the memory, then should they? It would be consistent with their autonomy to have the choice to take or refuse the drug. Each patient's decision would depend on how she weighed her experience of harm against the potential benefit and risk of the drug.

There is also the question of how a victim of an assault during surgery would weigh the benefit of erasing a memory of the assault against the social value of reporting the offender to the authorities for prosecution. This question should be informed by the high incidence of PTSD among patients experiencing AA, which may be higher in cases of assault. Some actions are more reprehensible than others. When a physician commits a criminal act such as sexual assault on a patient, would the patient be obligated to retain the memory to testify against the physician in a court of law? Or would she have the cognitive liberty to erase the memory as a way of preventing harm in addition to what she already has endured? These questions are complicated by the fact that memories are not always accurate reports of events. Recalling experiences is a reconstructive process. Each time a memory is retrieved, its content is updated to reflect one's actual circumstances. This can blur details of a past event and provide at best a fallible account of it. The emotionally charged content of an assault gives it salience and a vivid aspect. This can influence how the person describes the objective features of the event (Phelps, 2004). Combined with the tendency of episodic memories to become less detailed and more generalized over time, the emotional content of the memory may affect how one describes the event.

Confirmation that recall of a criminal act was accurate may be more likely if many patients reported being victims of the actions of the same physician. This occurred in the case of a Toronto anesthetist found guilty of sexually assaulting twenty-one female patients under conscious sedation during surgery (Jones, 2013). Even in these cases, though, if drug-induced retrograde amnesia were possible, then each patient should have the right to choose to erase or retain the memory of her experience. This involves broader legal issues pertinent to memory, which I discuss in Chapter 6.

Weighing the choices of taking or declining an experimental drug in cases of anesthesia awareness with recall would be complicated by uncertainty about the consequences of retaining a memory. It would also be complicated by uncertainty about the effects of a drug intended to remove it. These effects might include impairment of normal episodic and semantic memory processes and impaired cognitive and emotional functions. Nevertheless, the patient should have the right to make this decision, since she would have to live with its consequences. Given the limited time for memory erasure, the patient would have to quickly process complex information presented to her by the anesthetist soon after recovering consciousness. This could make it a fraught decision. Yet uncertainty about how an intervention might affect one's brain and mind does not warrant withholding medically relevant information from a patient for paternalistic reasons.

If a drug that could erase a traumatic memory were available, and if preliminary studies had shown that it was reasonably safe and effective, then taking the drug could be in the best interests of the patient with the memory. Provided that she was competent enough to weigh the potential benefit against the risk, respect for her autonomy would obligate the physician to disclose information about the drug to her. Stating that the drug might have unforeseeable side effects would not mean that the anesthetist was failing to discharge the duty to disclose information or was not acting in accord with the principle of nonmaleficence.

Data from studies of the effects of protein synthesis inhibitors and other drugs to modulate or erase memories of AA may not be available for some time. Clinical trials could clarify the benefits and risks. These trials would have to be designed to include patients reporting awareness shortly after surgery. They would have to be randomized and placebo-controlled to answer questions about the drugs' safety and efficacy. Knowing that the interval for modification was limited, a person with a traumatic memory would have to decide within a relatively short period after developing it if he wanted to participate in such a trial. A decision about whether to participate would not by itself determine the effects that an active intervention, or no intervention, would have on the patient's memory. Participating in the trial would only yield general knowledge about the potential of a drug to disrupt memory reconsolidation.

These hypothetical trials would be different from actual trials testing propranolol to prevent posttraumatic stress disorder in at least one respect. The purpose of the drug would be to erase the memory, not just attenuate its emotional content and leave its cognitive content intact (Pitman, Sander, Zusman et al., 2002). Eliminating the cognitive and emotional content of a traumatic memory by eliminating the memory trace would require stronger destabilizing effects on protein synthesis, synaptic connectivity and LTP. These effects might also entail some risk of affecting other, normal, episodic and emotional memories.

Implicit Memories

The type of memory involved in most discussions of anesthesia awareness with recall is explicit memory. This is memory that is accessible to conscious retrieval. Anesthetized patients may form other memories even when they remain unconsciousness throughout surgery. These are implicit memories and, as such, memories of which the patient is and remains unaware when she returns to consciousness (Mashour and Avidan, 2015, p. i24). Following Schacter's earlier research on implicit memory (1987, 1996, ch. 6), John Kihlstrom and Randall Cork explain: "Explicit memory is conscious recollection, as exemplified by the individual's ability to recall or recognize some past event. Implicit memory, by contrast, refers to any change in experience, thought or action that is attributable to a past event – for example, savings in relearning or priming effects" (2007, p. 635). Implicit memory "is unconscious memory" (p. 635). Priming is an implicit memory effect that occurs outside conscious recall.

As an example of priming, intraoperative suggestions by a surgeon or anesthetist are processed by unconscious mechanisms in the patient's brain and mind. Catherine Deeprose and Jackie Andrade distinguish between perceptual and conceptual priming (Deeprose and Andrade, 2006). Perceptual priming refers to the form of the stimulus, while conceptual priming refers to the meaning of the stimulus. The second type can be enhanced by semantic tasks. In their meta-analysis of priming studies, Deeprose and Andrade argue that perceptual but not conceptual priming can occur in the absence of conscious awareness under anesthesia (p. 1). Still, negative perceptual priming effects may result in psychological sequelae and adverse behavioral changes in people following surgery with general anesthesia.

Implicit memories are not factored into risk assessments of anesthesia awareness and recall because they develop from information outside awareness. This is due partly to the fact that traumatic memories resulting from AA are explicit memories of experience. It is also due to the difficulty in assessing the probability of a patient developing implicit memories and questions about whether subsequent behavior is evidence that a patient has them. Because the neural mechanisms mediating explicit memories are different from the mechanisms mediating implicit memories, drugs that can prevent, weaken or erase the first type may not have the same effects on the second. Midazolam, for example, could disrupt consolidation of an explicit memory but not disrupt consolidation of an implicit memory. Intraoperative suggestions may positively affect unconscious processing and result in improved psychological outcomes for patients. Depending on their form and content, however, these suggestions may also negatively affect processing and result in worse outcomes. Because this processing operates at an unconscious level, it can be difficult to predict whether a patient will develop these memories, or what their content will be. As Andrew Davidson points out, "you may not remember your colonoscopy, but you may become strangely uncomfortable when walking past a garden hose" (Davidson, 2014, p. 560).

A recent review of animal studies indicates the extent of implicit aversive memory under low doses of general anesthesia (Samuel, Taub, Paz et al., 2018). The underlying mechanisms involve several neurotransmitter systems in the hippocampus, amygdala and neocortex. A better understanding of these mechanisms might lead to preoperative or intraoperative interventions that might prevent or limit the extent of aversive conditioning in anesthetized patients. There are still significant knowledge gaps in the connection between anesthesia and implicit memory. Even if researchers could close these gaps, translating the results of aversive conditioning from animal to human models would be necessary to provide answers to questions about preventing harm from implicit memories.

One cannot update and modify implicit memories as one can update and modify explicit memories because the first type is not accessible to conscious retrieval. Priming before or during surgery could be one way of mitigating any deleterious effects of these memories on a patient's postoperative behavior. But careless use of words by a surgeon or anesthetist when a patient is unconscious may have a harmful impact on the patient's psyche. It is not known how or why implicit memories develop during the process that takes a patient to oblivion under general anesthesia and then returns her to consciousness. These considerations indicate that anesthetists have much less control of patients' implicit memories than they have of their explicit memories. Priming effects alone will not explain why some people but not others develop these memories, or the variable effects they have on their subsequent mental states and actions. Because of the greater degree of uncertainty about the formation of implicit memories, as well as uncertainty about their long-term effects on behavior, there would be no justification for an anesthetist to pre-operatively inform a patient about them. Doing so may only cause anxiety in the patient and be more likely to harm than benefit her. It could also thwart any potentially beneficial priming effects. Drawing the patient's conscious attention to the prospect of having these memories could interfere with the unconscious mechanisms in priming.

I have noted research indicating that anesthetics may have different effects on hippocampal and amygdalar mechanisms mediating episodic and fear memories. Anesthetics may prevent the first type of memories, but not the second. Veselis points out that there may be different mechanisms mediating explicit and implicit memories within the hippocampus itself. "It now appears that the hippocampus participates in processes independent of conscious awareness" (Veselis, 2015, p. i15). This may be part of a "more 'primitive' memory system than the human episodic memory system" (p. i16). Veselis defines this type of memory as "learning without awareness," or "subliminal learning" (p. i16). He asks, "What are the physiological underpinnings of learning without awareness, and if possible, where might these memories be stored? The latter question is difficult to answer because it is still not known 'where' conscious memories are stored, which may vary through time" (p. i16).

Consistent with Tulving's comments about storage of explicit memories cited in Chapter 1, Veselis's comments about storage of implicit memories suggest that they are not identical to or located in discrete sections of brain tissue. Instead, they are constituted by information associated with brain nuclei. Memories should not be described as entities but more accurately as informational properties that emerge from activity in distributed neural networks. Changes in explicit memories cannot be seen by looking directly in the brain but are inferred from changes in neural activation and synaptic connectivity based on brain imaging and changes in a person's behavior. Evidence of the formation of or changes in implicit memories is more uncertain because of incomplete knowledge of the neural correlates of this more primitive memory system. Postoperative behavioral manifestations of these memories may be ambiguous as well. This can make it difficult to know whether a patient formed implicit memories during AA. It indicates that inferences anesthetists or neurologists draw from behavior to these unconscious mental states are imperfect.

Most researchers agree that long-term declarative memories are stored in the neocortex, where information associated with these memories was first learned. Fear memories have cognitive representations in the neocortex and emotional representations in the amygdala. Because they are part of a more primitive memory system, one might think that implicit memories would be stored in subcortical regions. According to Squire and Zola's taxonomy, implicit memories associated with priming are stored in posterior regions of the neocortex. They may be mediated by a more distributed neural network consisting of subcortical, limbic and cortical regions. The constructive model attributes an adaptive purpose to explicit memories. Implicit memories may have an adaptive purpose independently of or in conjunction with explicit episodic, semantic and fear memories. Yet lack of conclusive evidence for implicit memories having such a purpose leaves this as an unconfirmed hypothesis. Whether these memories have beneficial or harmful effects on patients who have undergone general anesthesia may depend on preoperative and intraoperative suggestions from surgeons and anesthetists. There will always be some degree of uncertainty regarding these effects because researchers are far from completely understanding conscious and unconscious processing of memory at neural and mental levels.

Conclusion

Since William Morton's demonstration in 1846 that inhaled ether caused anesthesia and amnesia, research in anesthesia has highlighted the complex relationship between consciousness and memory (Shushruth, 2013). For a patient under general anesthesia, the boundary between consciousness and unconsciousness can be nebulous. It may not be clear during surgery where she is along the awareness spectrum. Because consciousness cannot be directly observed but only indirectly inferred from observation, imaging or electrophysiology, and because these techniques are fallible, it may not be known whether a patient experiences anesthesia awareness. A report from the patient after surgery may confirm that she was aware. This presupposes that she forms an explicit memory of awareness. Yet some patients have this experience and do not recall it, or they recall it long after surgery. Other patients form implicit memories of events while they were unconscious. These memories are inaccessible to conscious recall.

Preoperative infusion of propofol or benzodiazepines does not always prevent anesthesia awareness with explicit recall. Xenon or other gases or drugs with anesthetic and anterograde amnesic effects might contribute to a solution to this problem. But this would require additional research involving human subjects. Intraoperative infusion of these drugs would have to disrupt consolidation in progress to prevent an explicit memory from being formed and stored. Postoperative interventions aimed at erasing an explicit traumatic memory may not be effective if the memory has already been consolidated, reconsolidated and stored in the neocortex or amygdala. Attempts to induce retrograde amnesia may fail. Preventing these memories from consolidating or reconsolidating could prevent the harm they impose on people who have them. Critically, their efficacy would depend on when they were administered in the memory process. The only way of knowing their effects would be from patients' reports and observing their subsequent behavior.

In the introduction to this chapter, I cited a passage from authors claiming that anesthesia "can be a powerful tool for studying memory" (Kerssens and Alkire, 2010, p. 47). Yet we still have a poor understanding of how anesthetics work and how they affect consciousness and memory (Davidson, 2014, p. 560). Research in clinical neuroscience and cognitive psychology has confirmed *that* consciousness and memory have a neurobiological underpinning consisting of a distributed network of cortical, limbic and subcortical circuits. But the research has not satisfactorily explained *why* some people form and recall explicit memories of experiences and others forget them. Nor has it satisfactorily explained *why* some people form implicit memories and others do not. This applies both to the unique circumstances of intraoperative awareness in particular and to experience in general. The effects of explicit memories can be known from people's behavior. Yet how implicit memories affect behavior may be more subtle than apparent.

These considerations recommend caution in making claims about the connections between consciousness, anesthesia and memory. They also recommend caution in making claims about preventing or erasing memories. Anesthetics and sedatives can disrupt consolidation intraoperatively, but only if they are administered early in this process. While anesthetics, sedatives and benzodiazepines can prevent consolidation of episodic memories, they may not have the same effect on fear memories. They may disrupt explicit memory consolidation mechanisms in the hippocampus but not in the amygdala. The ability of these drugs to induce anterograde amnesia may be limited. They may have no effects on implicit memories. Erasing a memory that has been consolidated and is recalled days or weeks after surgery is still hypothetical. It is not clear whether a protein synthesis inhibitor or neuromodulation could induce retrograde amnesia.

If these interventions could safely and effectively prevent or treat sequelae from AA with recall, then one can ask whether they should also be used for those who suffer from the trauma of physical assault, natural disasters, motor vehicle accidents and other events. One can also ask whether nonpharmacological techniques could have the same effects. These interventions could be used as therapy for psychiatric disorders that may develop from memories of traumatic or disturbing experiences, including AA. The rationale for intervening in the brain and mind is to prevent or mitigate harm from these memories. This requires a more detailed neurophysiological explanation of how reconsolidation blockade and memory erasure are possible and the effects they could have on people with disorders of memory content. This is the focus of Chapter 4.

4   Disorders of Memory Content
and Interventions

Panic, phobia, PTSD, anxiety and some forms of depression are disorders of the fear memory system. Unlike disorders of memory *capacity*, such as anterograde and retrograde amnesia, they are disorders of memory *content* (Kopelman, 2002; Fradera and Kopelman, 2009). The problem is not an inability to remember experiences but an inability to forget them. More precisely, the problem is a persistent and intractable emotionally charged representation of a disturbing or traumatic experience. This can cause varying degrees of cognitive, emotional and volitional impairment in people who have it.

Disorders of memory content constitute a significant percentage of psychiatric disorders, including addiction. This involves disruption of mechanisms of learning and memory in pursuing rewards and the cues that predict them (Hyman, 2005). These disorders also include chronic pain, which can strengthen the emotional pathways that are activated during pain perception and remain activated when pain stimuli are absent (Apkarian, Baliki and Geha, 2009; Mansour, Farmer, Baliki et al., 2014). Information in the brain associated with nociception may remain stuck in the "on" position. The embedded memory can cause one to anticipate pain even when one is not experiencing it. Among what Schacter calls the "seven sins of memory," these disorders correspond to the "sin" of persistence (Schacter, 2001, ch. 7; 1996, ch. 7). They are involuntary sins because whether these memories develop and persist is largely beyond our conscious control.

In this chapter, I explain the etiology of disorders of memory content and why they are so often intractable. I consider different therapeutic interventions for them. Psychotherapeutic and behavioral techniques can attenuate the emotionally charged content of a pathological fear memory. Yet these techniques leave the cognitive representation of the memory intact. This may allow reactivation of its emotional representation by events reminding the subject of the initial experience that triggered the memory. Propranolol can also attenuate the emotional content of these memories. Here too, though, leaving the cognitive representation of a memory intact may allow reactivation of its emotional content. I consider other experimental interventions to erase memories. Protein synthesis inhibitors might block reconsolidation and erase pathological fear memories. High-frequency deep brain stimulation and high-intensity focused ultrasound may be more effective than drugs in erasing these memories because they can more directly target the nuclei constituting the memory trace. Not all fear memories are maladaptive, however. It is not known whether or how these drugs and techniques would influence adaptive fear memories and normal emotional, episodic and semantic memories. It is not clear how selective they could be in their modulating or erasing effects.

I discuss the potential effects of memory erasure on personal identity and authenticity. It would be desirable to erase memories with which we do not identify and which cause different types of impaired reasoning and decision-making. Pathological memories can disrupt the psychological connectedness and continuity that ground one's experience of persisting through time. I also discuss how erasing unpleasant but nonpathological memories might affect identity, rational and moral agency and authenticity. Not all episodic memories are necessary to sustain these psychological relations. One could erase some memories and edit parts of one's life narrative without undermining the unity and integrity of the whole. Nevertheless, some of these memories enable us to experience regret, to learn from our mistakes and plan more carefully for and act more effectively in the future. Other memories may cause us to dwell on the past and hinder our ability to form and execute action plans. Deleting some of them may enhance rational agency. Yet deleting memories necessary for generating and sustaining moral emotions like empathy and remorse could weaken one's moral sensibility and capacity to recognize and respond to the rights, needs and interests of others. Depending on the content and the actions they represent, a person could erase one or more memories and avoid these effects. Erasing selected memories may be consistent with authenticity if it results from a process of critical reflection on and endorsement of the mental states one wants to guide one's behavior.

The reconstructive model of memory supports the idea that one can retain identity and authenticity despite erasing pathological and some unpleasant memories. Just as we constantly update contextual information associated with memory to adapt to changing environments, we constantly shape our identities by processing information about these environments as we travel mentally from the past through the present to the future. Our identities are not fixed or complete by a certain age but always a work in progress. We do not discover our true selves but construct and reconstruct them as we go through life.

Pathophysiology of Fear Memories

The psychiatric disorders I have mentioned all result from hyperactivation of the fear memory system in the brain. In generalized anxiety disorder and some forms of depression, a persistent memory of a traumatic or disturbing experience distorts how one processes and interprets information about the world. It causes the affected person to perceive future events or situations as threats and with fearful anticipation. Excessive worry can result in avoidance behavior and impair the ability to form and execute action plans. Unpleasant memories of social interactions may contribute to social anxiety disorder, a subtype of generalized anxiety characterized by "an intense fear of social situations in which a person anticipates being evaluated negatively" by others (American Psychiatric Association, 2013, pp. 202–203; Leichsenring and Leweke, 2017). Panic disorder is a more severe form of anxiety disorder. It involves a more heightened emotional response to fear-inducing stimuli associated with a memory. Whereas panic disorder has a more acutely disabling effect on agency, anxiety disorder has a more chronically disabling effect.

There is more hyperactivation of the fear memory system in PTSD than in other psychiatric disorders of memory content. PTSD involves more persistent, emotionally heightened and intractable memories than anxiety. It also involves more extensive dysfunction in prefrontal and other cognition- and emotion-mediating brain regions to which the fear system projects. In addition to its debilitating psychological symptoms, PTSD results in more chronically disabling effects on agency. Approximately 20–30 percent of people who have experienced trauma also experience lifelong PTSD. It is the fourth most common psychiatric diagnosis (Parsons and Ressler, 2013).

In anxiety, panic, depression and especially PTSD, the stress of a traumatic experience can cause epigenetic changes in the brain. These changes alter gene expression and disrupt normal fear processing in the amygdala and other limbic circuits (Shalev, Liberzon and Marmar, 2017). There is dysregulation in DNA methylation and other mechanisms regulating transcription factors necessary for encoding, consolidating and storing memories. These are features of a pathogenic process that interferes with the normal transition from episodic memory formation in the hippocampal complex to long-term episodic memory storage in the neocortex. Emotionally arousing events activate the autonomic nervous system, which releases the stress hormone norepinephrine in the brain. In addition, the adrenal glands release glucocorticoid hormones like cortisol (Krugers, Zhou, Joels et al., 2011). Combined with the action of LTP and protein synthesis, these hormones embed the emotional information constituting the memory trace in the amygdala (Josselyn, 2010; Sillivan, Vaissiere and Miller, 2015; Van Marle, 2015; Zannas, Provencal and Binder, 2015). This can overstimulate the fear memory system, of which the amygdala is the most critical component. It disrupts normal cognitive-emotional processing mediated by a pathway linking the limbic system to the prefrontal cortex (PFC). This in turn interferes with the normal function of the PFC to inhibit emotional functions mediated by the limbic system. The emotionally charged content of the memory disrupts the brain's ability to process information internally from the body and externally from the environment. In PTSD, these pathological processes cause the intrusion of the traumatic memory in the form of flashbacks, nightmares, social avoidance and hyperarousal (Shalev, Liberzon and Marmar, 2017).

Joseph LeDoux distinguishes nonconscious from conscious aspects of fear (LeDoux, 2015, pp. 184–191). The nonconscious aspect pertains to the body's ability to detect and respond to threats to the organism. This involves activation of the amygdala circuit in the limbic system in the organism's response to external threats. It is distinct from the circuit in the brainstem that regulates fear and panic in the organism's response to internal threats from the body. The conscious aspect of fear is a product of cognitive systems in the neocortex. These systems project to and from the amygdala in processing fear at nonconscious and conscious levels. There is a cognitive representation of the memory in the neocortex and an emotional representation in the amygdala. The source of the psychopathology in disorders of memory content is not the cognitive representation of the memory but its emotionally charged representation. Theoretically, attenuating or erasing the memories causing or contributing to psychiatric disorders could prevent the intrusion of uncontrollable reminders of traumatic experiences in consciousness and fearful anticipation of the future. These interventions could weaken or remove the neural source of these psychopathologies.

Weakening and Erasing Fear Memories

Psychiatric disorders associated with a dysregulated fear memory system involve conditioned responses in which a memory of a traumatic event is reinforced by cues reminding the subject of that event. These cues cause the event to be repeatedly replayed in the subject's brain and mind. Research has demonstrated that memories can be modified in different ways. Cognitive and behavioral therapies aim to diminish emotional responses to stimuli perceived as aversive because of their association with the memory of a disturbing or traumatic event. One of these techniques is exposure therapy. The subject is gradually and repeatedly exposed to the event through cues that trigger conscious recall of it. This process can modulate the subject's psychological and behavioral response to the memory. The cognitive content of the memory is preserved. The subject's response can weaken its emotional content. Yet replaying a traumatic event in one's mind by repeatedly recalling it may reactivate this content and exacerbate rather than ameliorate one's emotional response to it.

The behavioral intervention of extinction training can attenuate conditioned fear responses to cues associated with the memory of the original experience (Kindt, Soeter and Vervliet, 2009). The subject learns to dissociate the memory of the experience from the cues. This can weaken the emotional content of the memory and the subject's response to it. The critical memory stage for this technique is retrieval, when the memory is labile and susceptible to change. Intervening behaviorally at this stage may prevent reconsolidation (Bjorkstrand, Agren, Ahs et al, 2016; Monfils and Holmes, 2018). Similarly, conditioning can allow fear memories to be updated with nonfearful information provided during retrieval and reconsolidation (Schiller, Monfils, Raio et al., 2010). This may eventually allow more recent normal episodic and semantic memories to replace older fear memories in the brain and mind's information store. Earlier studies suggested that behavioral techniques could enhance natural processes of neural systems suppressing unwanted memories (Anderson, Ochsner, Kuhl et al., 2004) A more recent study of intentional forgetting of visual experience showed that moderate activation of the engram made it vulnerable to disruption and weakening (Wang, Placek and Lewis-Peacock, 2019).

Behavioral techniques may be able to not just suppress these memories but gradually remove them from storage so that they do not compete with other memories for limited information space. These techniques could allow a transition from pathological or maladaptive to normal thought and behavior. Modulating epigenetic mechanisms involved in encoding, consolidating, retrieving and reconsolidating persistent fear memories could be one component of extinction training (Lattal and Wood, 2013). The effect of this technique would be to replace rather than erase these memories. This would be a safer form of memory modification than psychopharmacology or neuromodulation. Still, this would depend on whether the replacement was complete and eliminated the emotional content of the fear memory. Otherwise, certain stimuli could reactivate it.

The beta-adrenergic receptor antagonist propranolol can interfere with noradrenergic mechanisms associated with memory reconsolidation in PTSD. This can weaken the emotionally charged content of the memory while leaving its cognitive content intact (Pitman, Sanders, Zusman et al., 2002; Lonergan, Oliveira-Figueroa, Pitman et al., 2013; Villain,

Benkahoul, Drougard et al., 2016). The drug does not erase the traumatic memory but reduces the negative emotional valence associated with the remembered event. Results of studies testing this intervention in humans have been mixed. Researchers conducting a study concluded that "propranolol shows promise" in weakening powerful idiosyncratic emotional memories but emphasized that further studies are needed to confirm this effect (Lonergan, Oliveira-Figueroa, Pitman et al., 2013, p. 222). Even if this drug weakens these memories, the fact that the emotional trace of the memory is not removed leaves open the possibility that aversive stimuli could reactivate its emotional representation and the psychopathology. As Ryan Parsons and Kerry Ressler point out, when some emotional content of the memory remains, repeated retrieval limits the drug's action. This "can sensitize patients with fear-related disorders and lead to worsening of the disorder" (Parsons and Ressler, 2013, p. 151).

A more effective way of treating disorders associated with fear memories would be pharmacological blockade of memory reconsolidation. This would not just weaken the emotional content but eliminate the memory trace (Pitman, 2015). As I explained in the first three chapters, memories that have been consolidated need to be reconsolidated, or updated, to remain stored as information in the brain (Nader, Schafe and LeDoux, 2000; Nader and Einarsson, 2010; Nader, 2013). This occurs when they are retrieved in conscious or unconscious recall of the original experience. PTSD or other disorders could be treated by disrupting reconsolidation and removing the memory trace in the amygdala. Pharmacological intervention would occur during retrieval when memories are unstable and susceptible to alteration.

Roger Pitman explains: "For reconsolidation blockade, or updating, to be successful, two steps are required: First, the problematic memory must be destabilized. Second, its re-stabilization (reconsolidation) must then be prevented or modified (updated)" (Pitman, 2011, p. 2; 2015). Repeating a point from Chapter 3, because reconsolidation requires protein synthesis, a drug that blocked protein synthesis could block reconsolidation. Infusion of a protein synthesis inhibitor into the basolateral amygdala during memory retrieval could prevent reconsolidation and remove any trace of the memory (Gold, 2008; Agren, Engman, Frick et al., 2012; Parsons and Ressler, 2013; LeDoux, 2015, pp. 281–282, 301–309; Soeter and Kind, 2015). The drug would do this by interfering with LTP and CREB, both of which regulate protein synthesis and its effects on long-term memory. A protein synthesis inhibitor would not literally erase the memory by destroying neural tissue but would inactivate nuclei and excitatory synapses associated with the memory trace. "Nuclei are three-dimensional sets of neurons with their own unique cellular architecture and neurochemical identity" (Koch, 2012, p. 73). Functional imaging during retrieval of the fear memory could identify the hyperactive nuclei, and infusion of the drug could inactivate them. This would significantly reduce the possibility of reactivating the memory trace. There would be no possibility of reactivation if there were no longer any trace in the brain. The dose would have to be limited to avoid toxicity.

Blocking memory reconsolidation by blocking protein synthesis has been effective in disrupting alcohol-related memories in rats. Like other addictions, alcoholism is associated with persistent memories of cues that induce drug craving. A group of researchers identified areas in the amygdala and cortical regions sustaining these memories. By disrupting activation of mammalian target of rapamycin complex 1 (mTORC1) in these regions with a protein synthesis inhibitor, the researchers blocked reconsolidation of these memories and prevented relapse (Barak, Liu, Hamida et al., 2013). Consistent with the method of eliminating fear memories, blocking reconsolidation by inhibiting protein synthesis during retrieval can eliminate the memory driving addiction.

A different pharmacological intervention that could inhibit the formation of new memories and erase existing memories would be injecting zeta inhibitory peptide (ZIP) into the hippocampus. Microinjections of ZIP into rat hippocampi have had these effects. Initial studies suggested that ZIP could alter memories by blocking protein kinase M-zeta (PKM-zeta), which maintains synaptic connectivity in long-term memory formation (Sacktor, 2011). Yet more recent animal studies have shown that ZIP can disrupt LTP and other molecular processes necessary to maintain memories and erase them independently of PKM-zeta (Frankland and Josselyn, 2013; Lee, Kanter, Wang et al., 2013; Volk, Bachman, Johnson et al., 2013). The mechanisms of ZIP in disrupting the formation and maintenance of memories are not fully known. There is debate about the influence of genetics on these peptides and whether they enhance or inhibit memory formation (Sacktor and Hell, 2017). Despite the absence of trials testing the drug in humans, ZIP offers at the least the possibility of another pharmacological intervention in the brain to eliminate traumatic or unpleasant memories.

The anesthetic xenon is another potential treatment for PTSD and other psychiatric disorders. As noted in Chapter 3, xenon can inhibit reconsolidation of fear memories in rats. It can disrupt NMDA receptors to weaken fear conditioning associated with traumatic experiences. By blocking reconsolidation of recent memories, xenon could weaken the emotional representation of and possibly erase the problematic memory trace. Because xenon has both anesthetic and amnesic effects, it could be an ideal agent to prevent anesthesia awareness and recall. Yet because it influences distributed neural networks mediating awareness and memory, it is not known what its full neurological and psychological effects would be if it were used to disrupt reconsolidation.

For a drug to block reconsolidation and erase a memory, it would have to overcome some neurobiological obstacles. Parsons and Ressler note that "older and stronger memories are less susceptible to disruption after retrieval" and that "it may prove difficult to disrupt traumatic memories after retrieval because these memories are most certainly strong and, in many cases, have persisted for some time" (Parsons and Ressler, 2013, p. 151). Early intervention in the process of consolidation and reconsolidation would be critical. The studies I have cited confirm that reconsolidation blockade is more effective on recent memories and less effective on remote memories. This is especially critical for people with PTSD, many of whom may not recall or report a memory of a traumatic experience until weeks or months afterward. This could preclude effective treatment. Unless a reconsolidation-blocking drug is administered within a brief period after the experience, it may be too late to prevent the memory from becoming firmly embedded in the brain.

Another obstacle for pharmacological erasure of memory is selectivity, or specificity. Many memories of fearful experiences are adaptive and critical for survival. They enable us to recognize and respond appropriately to threatening situations. Not all fear memories are pathological or maladaptive. Because of the distributed and nondiscriminating effects of psychotropic drugs, a drug intended to erase a memory trace may have unintended expanding effects and impair normal function of the fear memory system. It may have similar effects on other declarative and nondeclarative memory systems. A drug infused into the brain could alter both targeted and nontargeted nuclei in the limbic system and alter normal emotional processing. This could introduce a new psychopathology. The selectivity problem of identifying molecular targets in fear and other memory systems and erasing particular memories applies to any drug that inhibits LTP and protein synthesis.

Neurostimulation of a hyperactive circuit in the fear memory system could overcome the problem of selectivity in erasing memories implicated in anxiety, depression, panic, phobia and PTSD. The technique involves more focused action in the brain than psychotropic drugs. Many neuroscientists describe neuromodulation techniques involving neurosurgery as invasive and those avoiding neurosurgery as noninvasive. This is a misleading distinction. Any technique that cause changes in neural activity is invasive, regardless of the location or source of the change. Accordingly, the distinction between invasive and noninvasive brain interventions should be replaced by the distinction between less invasive and more invasive interventions (Davis and van Koningsbruggen, 2013).

Less invasive forms of neuromodulation include transcranial magnetic stimulation (TMS), transcranial electrical stimulation (tES) and transcranial direct current stimulation (tDCS). In TMS, an electrical current passing through a coiled wire placed on the head creates a magnetic field that can penetrate the skull (Valero-Cabre, Amengual, Stengel et al., 2017). Magnetic pulses generate electrical potentials in the brain. These depolarize neurons and trigger action potentials in brain regions underlying the coils. The effects of a single pulse of TMS may last only a few milliseconds. Repeated pulses in repetitive transcranial magnetic stimulation (rTMS) may have more sustained effects. The general term "tCS" refers to techniques generating direct (tDCS) or alternating current (tACS) stimulation. In these techniques, the experimenter or practitioner typically delivers pulses of low-frequency electrical current for 10–20 minutes (Davis and van Koningsbruggen, 2013, p. 1). The main limitation of these techniques is that the pulses penetrate only the first centimeter of the cerebral cortex (Deng, Lisanby and Peterchev, 2013). They do not penetrate subcortical structures in the fear memory system and thus may be limited in modifying fear memories.

Deep brain stimulation (DBS) is among the more invasive neuromodulating techniques. This involves electrically stimulating dysfunctional neural circuits to modulate them so that they are neither metabolically overactive nor underactive (Hamani, Holtzheimer, Lozano et al., 2016). Using MRI-guided stereotactic techniques, electrodes are implanted unilaterally or bilaterally in a brain region identified as the source of neural and mental dysfunction. The electrodes are stimulated at varying frequencies through leads connected to a pulse generator implanted under the collarbone or in the abdomen. A programmable hand-held device controls the stimulation. The electrodes deliver intermittent or constant stimulation to modulate neural activity in the targeted area. DBS creates a virtual lesion by inducing electrophysiological silence in a neural circuit. Most of the circuits the technique targets are metabolically hyperactive, and its neuromodulating effect consists in downregulating this activity. DBS may be more effective than TMS or tCS in modifying hyperactive fear memories because it can penetrate subcortical structures implicated in their pathophysiology.

Magnetic resonance-guided focused ultrasound (MRgFUS) is a more recent form of neuromodulation. It uses acoustic waves transmitted through the skull from a device placed on the head to alter neurons and neural circuits. FUS can direct lower-intensity waves to modulate dysfunctional neural circuits or higher-intensity waves to ablate nodes within these circuits. FUS can penetrate deep brain structures mediating memory and other cognitive, emotional and motor functions without affecting neurons and tissue in higher brain regions. In contrast to DBS, FUS can penetrate and modify circuits in these structures without intra-cranial surgery (Lipsman, Schwartz, Huang et al., 2013; Elias, Lipsman, Onda et al., 2016). By avoiding the risk of infection and hemorrhage, FUS is a safer technique than DBS. It may not modify neural circuits as directly as DBS, however, because the acoustic waves travel through bone in the cranium. This could impede penetration of the waves to the neural targets and make it less effective than DBS.

In a different respect, FUS involves more risk than DBS. High-intensity FUS can ablate neural tissue by permanently destroying it. Any adverse neurological and psychological effects could be irreversible. Adverse effects of DBS could be reversible because the electrical frequency of the stimulation could be adjusted, the pulse generator could be turned off, or the electrodes could be surgically removed from the brain. In a different respect, FUS would be preferable to DBS because it avoids the question of how long to activate an implant in the brain following modification of the memory. As a less-invasive technique than DBS, FUS would not require implanting a device in the brain to weaken or remove the memory trace.

The more specific action of DBS or FUS on the neurons and synapses within the memory trace could avoid the limitations of protein synthesis inhibitors or other drugs. Direct stimulation of these constituents of the trace at the right frequency or intensity could neutralize the effects of CREB, LTP and protein synthesis on the persistence of the memory (Zhou, Won, Karlsson et al., 2009). By disrupting reconsolidation more directly, this could overcome any obstacles to destabilizing and removing the memory as stored information in the brain. In addition, by precisely targeting the neurons within the trace, stimulation could reduce the risk of expanding effects on adaptive fear and other emotional memories. Combined with its neuromodulating action, the ability of DBS to probe circuits would enable investigators to monitor its effects on critical neurons and synapses (Lozano and Lipsman, 2013). This could prevent adverse expanding effects on circuits unrelated to the problematic memory.

Lower-frequency DBS or lower-intensity FUS may be limited in only attenuating the emotional content of the memory trace. This effect would be similar to that of propranolol because it would not eliminate this content. Higher-frequency DBS or higher-intensity FUS could selectively inactivate neurons and synapses in the basolateral amygdala identified as the source of the pathological fear memory, disrupt reconsolidation and erase it. Still, fear, pain and addiction involve learned behavior. Inactivating neural circuits alone may not be sufficient to produce the desired effect. Behavioral interventions may be necessary as well.

It is possible that ablated nuclei could become reinnervated. Depending on how the subject and his brain responded to stimuli, there could be a reactivation of the nuclei and the memory trace. This has occurred in some cases where lesioned hyperactive circuits associated with obsessive-compulsive disorder returned to their hyperactive state (De Ridder, Vanneste, Gillett et al., 2016). The memory might not be erased after all. This may indicate the need for continuous rather than intermittent stimulation of the targeted brain region.

These techniques might also be able to erase a memory trace associated with hyperactive nociceptive pathways in chronic pain (Sandkuhler and Lee, 2013; Glannon, 2018). Unlike radiofrequency or high-intensity FUS ablation, a DBS-induced lesion would only be virtual. DBS could remove the memory by inactivating the nuclei without destroying any neural tissue. Consistent with Tulving's theory of the ontological status of memories, "erasing" does not mean removing an object in the brain identified as the memory. Rather, it means disrupting information in nuclei from which the memory emerges as a property of that information. Through its disruptive action on the information, DBS could remove this property from the brain and mind.

The selectivity problem is a problem about localization. The main question is whether a pathological fear memory would be localized enough for DBS or a different technique to erase it and leave other memories intact. Functional imaging could reveal high levels of activation in the nuclei associated with the memory when the subject was asked to recall it. This could confirm specific neurons and excitatory synapses in the amygdala as its source (Pitman, 2015). DBS could then target and inactivate these nuclei to erase the memory trace. It would do this directly by stimulating the nuclei at the right frequency and for the necessary duration rather than indirectly by blocking reconsolidation. PET or fMRI could confirm this effect by showing differences in glucose metabolism and blood flow corresponding to differences in neural activity and synaptic connectivity before and after the technique.

Despite the known probing and modulating effects of DBS, its mechanism of action is not well understood. It is not known whether inactivating nuclei in a discrete region of the brain would have expanding effects on normally functioning circuits mediating other memories. The aim of this intervention would be to permanently remove the memory trace from storage in the amygdala. Yet any irreversible effects on neurons and synapses mediating normal fear, positive emotional and spatial memories could harm persons by impairing their ability to navigate the environment. It is also possible that DBS or FUS could weaken or erase episodic memories and disrupt the psychological connections constituting one's autobiography. Behavioral interventions that replaced fear memories with normal episodic memories rather than attenuated or ablated them might avoid these problems. As noted, though, they may not eliminate the possibility of stimuli reactivating the emotional content of the memory. Ablative techniques may not be able to prevent reinnervation of neurons associated with the memory trace. These interventions may not permanently erase the problematic memory.

Researchers conducting clinical trials testing these interventions would have to obtain informed consent from mentally competent subjects and provide adequate protection in not exposing them to more than minimal risk. They would be discharging the bioethical principles of respect for personal autonomy and nonmaleficence (Beauchamp and Childress, 2012, chs. 4 and 5). In people with severe psychiatric disorders, cognitive impairment may preclude the informed consent necessary to participate in these trials. Whether they could participate would require a psychiatric assessment of their cognitive and emotional capacity.

Another example of the selectivity problem is that stimulation of the amygdala for PTSD may induce a nonspecific emotional response, such as crying, in a patient without any external stimulus. This may have no connection to a fear memory. These and other potential neural and mental effects of DBS need to be considered before using it to modify memories. It is also important to emphasize that emotions are mediated not just by the amygdala but by a circuit consisting of interacting cortical-limbic pathways. For example, the amygdala and the orbitofrontal cortex (OFC) together mediate the experience of regret (Camille, 2004). An intervention targeting a circuit associated with a fear memory could affect the amygdala's projections to the OFC and impair the capacity for regret or other emotions.

Optogenetics is a more recent technique that might be used to treat disorders of memory content. This technique uses light-sensitive proteins (opsins) that conduct electricity to control activity in neurons and neural tissue. Neurons are genetically modified to express light-sensitive ion channels. Exposing neurons to light can activate or inhibit electrical activity in the brain (Deisseroth, Etkin and Malenka, 2015). In optogenetics, inserted genes are not disease-related but instead genes that make neurons light-sensitive. They could be inserted in brain regions with different degrees of neural dysfunction. Optogenetics has better spatial resolution and is a more focused means of neuromodulation than DBS or FUS. It also has high temporal resolution, making it superior to the delayed and distributed effects of psychopharmacology.

One potential application of optogenetics would be to erase fear memories associated with PTSD, anxiety or panic disorders. In one study, researchers used this technique to weaken a fear memory in mice (Kim and Cho, 2017).

The researchers activated the fear memory in the amygdala by generating a high-pitched tone. They identified the memory trace in this brain region by observing activation in the neurons and synaptic connections constituting the trace as the mice responded to the sensory cue. The light selectively stimulated and inactivated only the neurons and synapses associated with the trace. Greater neuromodulating precision of optogenetics compared with protein synthesis inhibitors, DBS and FUS could overcome the problems of localization and specificity. If its effects were highly specific, then it could erase a maladaptive or pathological memory while preserving normal adaptive fear and episodic memories. In a different study involving a transgenic mouse line, optogenetic stimulation of neurons associated with extinction memories suppressed neurons associated with competing fear memories of place in the hippocampal dentate gyrus (Lacagnina, Brockway, Crovetti et al., 2019). This suggests the possibility of permanently silencing the problematic neurons and preventing reactivation.

The ability of high-frequency DBS, high-intensity FUS or optogenetics to inactivate or eliminate a memory trace would be preferable to protein synthesis inhibitors used for this same purpose. These techniques inactivate neurons and excitatory synapses constituting the pathological fear memory in the amygdala more directly and precisely than psychotropic drugs. As more focused means of neuromodulation targeting circuits in specific brain regions, the techniques could avoid many of the distributed adverse effects of these drugs.

The potential benefit of these neuromodulating techniques would have to be weighed against the risk, however. Risk is significant in techniques involving brain implants, including DBS and optogenetics. These are highly invasive procedures requiring intracranial surgery to place and activate devices in the brain. Risk is proportional to invasiveness. The greater the degree of invasiveness, the greater the risk. Implanting devices could cause bleeding resulting in neurological and psychological sequelae. Activation of the electrodes in DBS could affect both dysfunctional and normally functioning neural circuits. The same potential for unintended expanding activation would apply to the fiber-optic light source in optogenetics. This intervention would have the double risk of overstimulating neural circuits and causing unanticipated events from manipulating genes in the brain. Even if these techniques avoided these effects, there would still be the separate risk of bleeding and infection from surgically implanting devices in the brain.

Given these risks, the problematic memory would have to be refractory to all other interventions. The probability of erasing it would also have to be high to justify using these techniques. Protein synthesis inhibitors would appear to involve fewer risks and be safer than these other interventions. Yet they may not be as effective. Even if all forms of memory erasure were neurobiologically feasible, there would be trade-offs between their safety and efficacy that would have to be weighed in selecting among them. Electroconvulsive therapy (ECT) is a good illustration of how any intervention in the brain involves trade-offs. Many patients receiving ECT for unipolar and bipolar depression experience improved mood and motivation. But a significant number also experience some degree of retrograde amnesia. Some studies suggest that some memory loss could be associated with the therapeutic effect of ECT. In a study involving patients with unipolar depression, this technique disrupted reconsolidation of emotional episodic memories by disrupting memory retrieval (Kroes, Tendolkar, van Wingen et al., 2014). Reports of amnesia from patients undergoing the procedure typically involve normal episodic memory, and this effect is not therapeutic. Nothing in the study indicated how ECT might weaken fear memories while leaving other memories intact. The induced generalized seizures can ameliorate depressive symptoms. Yet in too many cases, this benefit comes with some loss of episodic memory.

To date, the use of protein synthesis inhibitors, xenon gas and optogenetics to disrupt reconsolidation and erase memories has been limited to animal models. We can only speculate on how these interventions would influence memory in humans. Most studies using DBS to modify memory in humans have been designed to enhance or improve it, and I will discuss some of them in Chapter 5. Some studies have shown that stimulation of a circuit in the medial temporal lobe can disrupt episodic memory consolidation and reconsolidation in humans (Fell, Staresina, DoLam et al., 2013; Merkow, Burke, Ramayya et al., 2017). It is unclear whether high-frequency electrical stimulation of the basolateral amygdala would have the same effect for people with pathological fear memories. Research into the use of neurostimulation to attenuate or erase memories is still at an early stage. Placebo-controlled clinical trials would be needed to establish that these techniques could have these effects. Accordingly, we must be circumspect in making claims about erasing memories. Nevertheless, there is an empirical basis for this action in humans. One group of researchers found that cigarette smokers with brain damage to the insula, which is associated with conscious urges, were more likely to stop smoking than those without this damage. The injury caused the smokers to "forget" their addiction (Naqvi, Rudrauf, Damasio et al., 2007). Inactivating nuclei in the memory trace of a hyperactive insula with DBS or FUS may be one way of treating refractory addiction.

Inactivating a persistent memory trace associated with chronic pain may be more challenging. Pain perception involves a more distributed neural circuitry. As I explained in Chapter 3, there is a lateral pain system, or sensory network, and a medial pain system, or affective network (Demertzi and Laureys, 2012). These networks involve interaction between many brain regions. While fear memories may be localized in nuclei in the amygdala, there is no discrete pain center in the brain. Despite the suggestion that targeting the insula might resolve addiction, there is no single neural source of pain that could be targeted to treat it. The distributed neural underpinning of pain would complicate attempts to use protein synthesis inhibitors or neuromodulating techniques to erase pain memories.

Like addiction, chronic pain is to some extent learned behavior. For this reason, cognitive behavioral therapy (CBT) techniques may be more effective for this condition than for pathological fear memories. These techniques may enable a subject to unlearn the connection between cues from stimuli and the memory of pain. This could alleviate pain, or at least diminish the expectation of it. Like behavioral therapies that update fear memories with nonfearful information, CBT that updated information associated with pain memories and gradually replaced it with new information might weaken or even remove the memory trace from the nociceptive network. These interventions might relieve pain noninvasively by modulating neural and mental processes involved in conditioning and expectation. Yet just as weakening or erasing pathological fear memories would have to be selective enough to avoid weakening or erasing adaptive fear memories, weakening outsized chronic pain memories would have to be selective enough to avoid weakening normal pain perception. Expanding effects of memory modification on normal as well as abnormal pain processing could cause more harm than benefit to patients. Noninvasive therapies like CBT would be less likely than neuromodulation to have these effects. As with other interventions, though, they would leave open the possibility of stimuli reactivating the pain memory trace.

Suppose that investigators could use psychotropic drugs, electrical stimulation or focused ultrasound to selectively and safely erase not just pathological but also less emotionally charged memories of unpleasant experiences. Although they do not necessarily cause maladaptive behavior, these experiences may impose constraints on our thought and behavior. We may be better off forgetting them. But do unpleasant memories serve a prudential and moral purpose? Would erasing them be permissible? Could the potential benefit of erasing memories that do not significantly impair one's cognitive, emotional and volitional capacities justify the neurological and psychological risk?

Memory Modification: Effects on Identity and Agency

Pathological traumatic memories and nonpathological unpleasant memories are different types of memory. The first type involves hyperactivity of the fear memory system that significantly impairs cognitive and emotional capacities. The second type involves some degree of anxiety or distress without the same degree of cognitive and emotional impairment. Although my discussion of weakening and erasing memories with drugs or neuromodulation addressed pathological memories, these same interventions could be used to weaken or erase unpleasant memories.

In a 2003 working paper, the US President's Council on Bioethics expressed concern that pharmacological modification of memory could have untoward consequences in our lives (President's Council on Bioethics, 2003; Kass, 2003). The Council worried that dampening or erasing memories might erode our moral sensibility. Emotions such as shame and regret depend on episodic memories of our actions and omissions. Altering memories could erode these emotions and remove psychological constraints on doing things we would not otherwise do. Given the essential role of episodic memory in our conscious lives, altering memories could reshape our lives in ways that might not be beneficial or benign. The Council stated: "As the power to transform our natural powers increases both in magnitude and refinement, so does the possibility for 'self-alienation' – for losing, confounding or abandoning our identity" (Kass, 2003, p. 294).

Although the Council did not clearly distinguish traumatic from non-traumatic memories, it acknowledged that some memories are so severe that they can destroy the lives of the people who suffer from them. Pharmacological or other interventions that weakened or erased these memories could be justified. When PTSD or other disorders of memory content severely affect a person's ability to function, there may be compelling reasons to use propranolol, a protein synthesis inhibitor, or neuromodulation to weaken or remove the memory from the brain and mind. Given the magnitude of harm from these memories, these interventions could be justified despite the risk of altering normal episodic memories or other neurological and psychological sequelae. The potential benefit from neutralizing a problematic memory could outweigh the risk.

Losing some of the emotional content of normal episodic memories may be an acceptable price to pay for altering a traumatic memory. But it would be acceptable only if the memory had an outsized negative impact on the person's mental states and adversely affected her quality of life. Unlike the pathological fear memory in PTSD, this proviso would make it more difficult to modify memories that were unpleasant but not pathological. Weakening their emotional representation could impair our ability to understand how our memories fit together in a meaningful narrative. Depending on the extent to which memories were modified, one might be left with the capacity to recall events without understanding how they were connected. Removing the emotional content of episodic memories could transform them into isolated semantic memories devoid of any personal meaning. Features of the natural and social contexts in which we experience events constitute the content of the memories we form and retain of them. This content is also a critical component of meaningful recall of our experiences. The connection between context and content in episodic memory forms the spatial and temporal structure of our autobiographies (Medford, Phillips, Brierly et al., 2005). Removing the emotional content of these memories, or removing the memory trace altogether, could disrupt the narrative unity and continuity of our mental life. Because the prospect of memory modification is hypothetical, these concerns are to some extent speculative. But it is instructive to consider the possible psychological effects of altering unpleasant memories.

As I discussed in Chapter 2, memory is a fundamental component of the psychological connectedness and continuity in terms of which many philosophers define personal identity. Many use "identity" and "self" interchangeably, and I follow this practice here (Rowlands, 2017). Episodic memory is critical to maintain these psychological relations. By enabling one to reexperience the past and anticipate the future, it is essential for the experience of persisting through time as the same individual. Presumably, erasing one or more episodic memories would disrupt psychological connectedness and continuity and thus identity. Memories are especially important in their forward-looking function of simulating events and guiding our actions. As part of this function, memories must be updated to make them relevant to current and anticipated situations. Updating is necessary for the flexible thought and behavior that enables us to adapt to changing environments. This may involve some degree of misremembering past events. It may also involve the replacement of information associated with older memories by information associated with more recent memories. If updating episodic memories is part of the process of maintaining the psychological relations of identity in a normal lifespan, and if updating means losing some of these memories, then one may lose some episodic memories without losing one's identity. Not every memory is necessary to sustain personal identity. Erasing some of them may be compatible with or even enhance the adaptive purpose of memory.

Identity is closely linked to agency. The experience of mental time travel involves forming action plans at earlier times and executing them at later times. Losing some episodic memories naturally or artificially could be a way of preventing the brain from having to process too much information, which could impair reasoning and decision-making. Eliminating some episodic memories that are not necessary for agency, and especially those that interfere with it, may improve the function of other episodic memories in promoting goal-directed behavior. This would be an example of a diminishing concept of cognitive enhancement (Earp, Sandberg, Kahane et al., 2014). The set of memories is diminished to allow for more efficient information processing at a brain-systems level and more efficient and effective reasoning and decision-making at a mental-systems level. Sometimes having less is better than having more. Too many memories can clutter the brain and mind with irrelevant information and impede the exercise of these capacities. Information overload can be particularly disabling when it consists of fear and disturbing memories.

Personal identity requires a continuous set of unified and integrated memories. Some memories naturally replace others, and we revise the content of memories to make them relevant to circumstances in our lives. But provided that there is a stable core of memories that persists through these changes, identity can be preserved. Profound amnesia undermines identity because the loss of many episodic memories undermines this core. In many cases, though, one could lose some memories while leaving the core intact. Conceiving of identity as a narrative constituted partly by memories, we can edit some memories in this narrative without affecting the unity and integrity of the whole (Schechtman, 2014, ch. 4).

Consistent with the reconstructive model of episodic memory, identity is a work in progress. The psychological relations that form our identities are not fixed and complete by a certain age. We constantly shape and reshape them in an open-ended process. Most of us have no memories of our experiences before age three or four. The narratives that we construct from episodic memories in our lives are not well developed until late adolescence or even early adulthood. These narratives are based on a scaffolding of these experiences. The memories that form this scaffolding need to be updated so that one's narrative aligns with the actual world. By updating the content of these memories and how this process influences our perception of the past and future, we are constantly constructing and reconstructing our selves. Amnesia consisting in the loss of many episodic memories could cause a collapse of the mental scaffolding that sustains the self. Erasing a significant number of memories could have a similar effect.

The goal of memory erasure would be to eliminate only memories that were incongruent with the mental states we wanted to have as the basis of our mental time travel and the source of our actions. Erasing one or a few memories would not weaken the psychological relations necessary to maintain identity. Nor would it undermine the set of motivational states necessary for effective agency in meeting the cognitive and emotional demands of the natural and social environment.

Nina Strohminger and Shaun Nichols argue that moral traits and how they affect social relationships are the most essential part of personal identity. These traits are more critical to identity than memory (Strohminger and Nichols, 2014, 2015). Our selves are more of a reflection of our moral character than our capacity to recall and use information about the past to project ourselves into the future. Our true selves are moral selves. The case of Phineas Gage illustrates this point. Gage lost many of his rational and moral capacities from damage to his ventromedial prefrontal cortex when a tamping iron penetrated this region of his brain in an explosion while he was working on the Rutland & Burlington Railroad in Vermont in 1848. The changes in Gage's personality and behavior were so significant that, "in the words of his friends and acquaintances, 'Gage was no longer Gage'" (H. Damasio, Grabowski, Frank et al., 1994, p. 1102). There have been conflicting accounts of the effects of the injury on Gage's long-term behavior (Harlow, 1848, 1868; Bigelow, 1850). Gage reportedly recovered many of his mental functions and became a stagecoach driver in Chile. But the initial period following the injury appeared to confirm the role of moral traits in the judgment that he was not the same person after the accident.

There are first-person subjective and third-person objective aspects of personal identity. The first aspect pertains to one's experience of mental time travel from the past through the present to the future. The second aspect pertains to how others perceive the person based on her behavior, as in the comment that "Gage was no longer Gage." Although episodic memory may not be essential to the second, objective aspect of identity, it is essential to the first, subjective aspect. It grounds the phenomenology of diachronic autonoetic awareness at the core of the self. A patient with unipolar or bipolar depression who loses some episodic memories in retrograde amnesia as a side effect of ECT may report the experience of a temporal gap in the connections between her past and present mental states. Depending on the extent of the amnesia, she may experience a disruption in her subjective sense of identity. This has nothing to do with moral traits and everything to do with memory. Fugue states (dissociative fugue) and other forms of TPA can cause one to lose one's orientation in space and time and forget one's autobiographical properties. As noted in Chapter 2, people with these disorders "lose their personal identity" (Goldstein and Kapur, 2012, p. 273), if only temporarily. Profound anterograde and retrograde amnesia caused H.M. and Clive Wearing to experience mainly fragmented episodes of their lives in the present tense. In all of these cases, the disruptive effect of these states on identity has nothing to do with moral traits and everything to do with memory.

Alzheimer's disease and other dementias involve substantial changes in both first- and third-person aspects of identity. Deterioration in cognitive, emotional and volitional functions corresponds to deterioration in episodic, semantic, working, spatial and prospective memory. Changes in personality and loss of rational and moral capacities in these disorders are due to degeneration of a range of memory functions resulting from degeneration in multiple regions of the brain. Alzheimer's patients forget how to inhibit their behavior and how to act in accord with rational and moral norms. In advanced stages of the disease, they lose their sense of space and time. This neurodegenerative disease shows that declarative memory plays a critical role in both self-awareness and the behavior through which others perceive us.

The ethical implications of manipulating human memories are equally significant (Glannon, 2006, 2011; Liao and Sandberg, 2008; Elsey and Kindt, 2016; Lavazza, 2016; Liao, Sandberg and Savulescu, 2016). Although erasing one or a few episodic memories would not necessarily disrupt personal identity, it could have positive or negative effects on self-regarding and other-regarding behavior. This is where the moral trait model defended by Strohminger and Nichols becomes important because it highlights the connection between identity and agency. If the emotional content of an episodic memory has a stronger influence on one's actions than its cognitive content, then taking a drug such as propranolol to take the emotional sting out of a memory could have the same impact on rational and moral agency as erasing the memory (Henry, Fishman and Youngner, 2007; Kolber 2006). Whether retaining or eliminating this content promoted or impeded our action plans would depend on the emotional valence of the memory, the experience it represents and how one responds to it.

Unpleasant memories do not cause the same degree of cognitive and emotional impairment as traumatic memories. Still, they may interfere with forming and executing action plans. Frequent recall of mistaken choices and actions can haunt a person for years. It can cause one to become overly conscious of these mistakes and stuck in the past. This can thwart the capacity to imagine future events and cause one to be tentative and indecisive. Ridding oneself of these memories could make one more assertive and resolute in acting. Yet these memories may also promote conscious reflection and deliberation resulting in more prudent choices. In addition, memories of unpleasant or disturbing experiences are necessary for the moral emotions of empathy and remorse that can make us more responsive to the rights, needs and interests of others. They are essential for moral growth and the development of moral character. Retaining memories of being wronged or harmed is also necessary to forgive those who wronged or harmed us. Forgiveness would require a weakening of the negative emotional content in recalling these experiences. But one would have to retain some of this content and the cognitive content of the memory to forgive them.

Some degree of negative emotional valence in unpleasant memories may be necessary to learn from past mistakes, gain insight into our behavior and modify it accordingly. It may be necessary for empathy and other social emotions like regret, shame and remorse associated with moral character and the capacity to respond to reasons for and against different actions. This too can be part of adaptive thought and behavior because interaction with others is one aspect of an adaptive response to the social environment. Eliminating episodic memories from our brains and minds could thwart these processes and impair the capacity for effective rational and moral agency. Neuromodulation to weaken or erase unpleasant memories may target a frontal-limbic circuit involving the OFC. This could affect not only our moral but also our prudential reasoning. In addition to mediating the emotion of regret, this region is critical for counterfactual reasoning in choosing between different courses of action (Van Hoeck, Watson and Barbey, 2015).

Tinkering with some of these memories would not necessarily erode one's moral sensibility. Weakening the emotional content or erasing one or a few of these memories could benefit a person without weakening her capacity to recognize and respond to self-regarding prudential reasons and other-regarding moral reasons. Depending on the action or omission one recalls, and the extent to which recalling it affects one's mental life, therapeutic forgetting may be justified in some cases.

Suppose that a young scientist gives a lecture at a major conference. The data he presents includes an egregious factual error, and some members of the audience harshly criticize him for it. He fails miserably in trying to respond to their criticisms. He is embarrassed by this experience. It causes him to lose self-confidence and to question his ability to do scientific research. The memory is not so emotionally charged that it puts him at risk of developing a psychiatric disorder. But it is disturbing enough to cause him to suspend his research. Taking a protein synthesis inhibitor or other reconsolidation-blocking drug that safely and effectively erased the memory could be defensible in this case. It would prevent additional psychological harm from the memory and benefit the scientist by causing him to forget the episode, remove the mental constraint it imposes on him and enable him to resume his research. This would be a case of justified therapeutic forgetting. Modifying the memory of an act that involved no wrongdoing would not necessarily impair his capacity for moral emotions and to recognize and respond to moral reasons. It would not necessarily incline him to perform harmful acts that he would not otherwise perform. Erasing the memory would not cause him to become imprudent or immoral (Glannon, 2011).

In contrast, a person who erased a memory of aiding and abetting war crimes, or a memory of an assault he committed, could weaken his capacity for moral sensibility. Erasing the memory could desensitize him to the wrongfulness of his action and the potentially harmful effects of his future actions on others. In cases where one completely lacked the capacity for moral sensibility and moral agency, retaining or erasing a memory of a harmful act might not make any difference to his subsequent behavior. Erasing the memory would not serve a rehabilitative or deterrent purpose. This severe type of cognitive and emotional deficiency could be the result of traumatic or fear memories of harmful acts the person experienced from others. The persistence of these memories could partly explain the person's actions and why removing or retaining his memories of them might not influence his future actions.

Extinguishing a memory of a harmful or wrongful act could impair a person's responsiveness to social norms and the capacity for moral behavior. It could eliminate the mental states that would be the motivation for changing his behavior. Deliberately removing one or more of these memories of these actions could also cause a substantial change in both first- and third-person aspects of his identity. It could cause a substantial change in the content of his mental states and how it affected his overall experience. It could cause a substantial change in how others perceived him and judged his behavior as well. Nevertheless, this would not be a likely consequence of an experience in which only the subject was affected by his action or omission, and he did not harm or wrong another person. Erasing the memory would not cause a substantial change in his moral character or identity. On the contrary, it could enhance his agency by removing a mental obstacle to practical and moral reasoning and decision-making.

There may be prudential and moral reasons for erasing memories when they cause psychological harm to the person who has them. Because of the difference in the extent of harm to oneself, the reasons for erasing traumatic memories are stronger than the reasons for erasing unpleasant memories. Reasons for erasing episodic memories are weak, or even indefensible, when the action the memory represents causes harm to others. The magnitude of the harm would provide a decisive reason for retaining the memory to make the agent appreciate the wrongfulness of his action. Except in cases where one is severely deficient in or lacks moral sensibility, retaining the memory of a harmful act would be necessary to generate and sustain moral emotions such as remorse. These emotions could lead one to recognize the wrongfulness of the action and possibly make amends for it. Such a response could also influence predictions about whether or to what extent a criminal offender could be rehabilitated. It is important to emphasize that it is not only the cognitive trace but also the emotional trace that is critical for the moral and legal significance of memory.

Memory Modification and Authenticity

Even if weakening or erasing memories had only salutary effects on prudential and moral behavior, some might be concerned that memory modification would conflict with the idea of authenticity. They might be concerned that tinkering with memories would threaten our true selves. According to Charles Taylor, authenticity is a moral ideal. "Each of us has an original way of being human," and a "measure" that determines what counts as one's "own way" (Taylor, 1992, pp. 28–29). Being authentic means that it is up to each of us to find our own way in the world, to flourish and to be true to our self. "If I am not true to my self, I miss the point of my life, I miss what being human is for me" (p. 29). Presumably, being "true" to oneself and thus authentic means acting in a way that is consistent with one's considered interests and values. These psychological properties ideally align with the desires, beliefs and intentions that issue in actions.

Authenticity presupposes autonomy, whereby one's actions and the motivational states from which they issue are one's own. This involves a hierarchical process of critically reflecting on and identifying with, or endorsing, certain mental states as the genuine source of one's actions (Frankfurt, 1988a, 1988b, 1988c; G. Dworkin, 1988; Mele, 1995). Through this reflection, a person can ensure that her first-order motivational states reflect her second-order attitudes about the types of states she wants to move her to act. Identifying with one's motivational states and actions makes them "true" to one's self and thus autonomous and authentic.

Would erasing one or more unpleasant memories be compatible or incompatible with authenticity (Erler, 2011)? Intuitively, losing a memory of an unpleasant experience would not alter one's interests and values. But if this memory were critical for prudential and moral growth, then losing it could weaken these essential dimensions of the self. The President's Council was concerned that manipulating memories could result in self-alienation. Erasing episodic memories could conflict with our interests and values and alienate us from our true selves.

The concept of authenticity has figured prominently in the debate on cognitive enhancement. One of the main questions in this debate is how enhancing cognitive functions beyond normal levels affects one's true self. The permissibility of enhancement depends to some extent on whether one defines authenticity as self-discovery or self-creation. It depends on whether the self is something inherent in our nature that underlies and is reflected in our actions, or something that we construct from our actions. It also depends on how drugs or techniques that alter memory affect the self (Taylor, 1992; DeGrazia, 2000, 2005). Differences between these two definitions of authenticity and the self revolve around what Taylor calls the "measure" of one's own way. Specifically, it revolves around the question of whether this measure is a property that we find within us or a property that we construct with our autobiographies through the different stages of our lives.

Is what Taylor calls the "inner idea" or the "true self" (1992, p. 29) an innate feature of our human nature? Or is it a feature of our psychology that emerges from our actions and life experience? If the true self emerges from our actions, then is there a point beyond which it should not be tinkered with? What role does memory have in the concept of an authentic self? How does weakening the cognitive and emotional content of a memory, or erasing it, affect authenticity? These questions need to be addressed regardless of whether memory modification is described as enhancement or therapy.

Following Taylor, a proponent of authenticity as self-discovery, Carl Elliott, says that "the ethic of authenticity tells us that meaning is not to be found by looking outside ourselves, but by looking inward. The meaningful life is an authentic life, and authenticity can be discovered only through an inner journey" (Elliott, 2003, p. 35). Further, he says that the "true self is the one that sits alone, a solitary self that endures over time, while the onstage self is a mere persona, a type of useful role-playing that can be used or discarded as circumstances demand" (p. 3).

It seems plausible to define authenticity as a process that we initiate and sustain through our actions rather than an intrinsic property that we recognize within us. Being authentic means acting in accord with one's considered interests and values. These are not fixed and unalterable but dispositions that we develop and revise to meet the cognitive demands and expectations of the natural and social environment. Authenticity is not discovered through an entirely inner journey. It is constructed through interaction with other persons and the world. Authenticity is the product of factors both internal and external to the person, and updating memories is one aspect of becoming authentic. We use and revise information about the past to respond to changing circumstances. If retaining a memory causes us to focus too much on the past and impairs the capacity for flexible and adaptive behavior, then erasing the memory can remove this impairment and promote this behavior. In this respect, erasing a memory can be an authentic act.

This does not imply that our selves are entirely self-created. The memories in the psychological relations constituting the self are to some extent beyond our control. We do not create memories or the processes through which they become encoded, consolidated, stored and reconsolidated in the brain and mind. The reconstructive model of memory consists in the process of updating its cognitive and emotional content as a framework for current and future action. We construct meaning from memories and develop our selves through this updating process. This *constructive* concept is distinct from the *creative* concept of the self and meaning in existentialism. For example, Sartre writes that a person is "nothing other than what he makes of himself" (Sartre, 2007, p. 18). The structures of meaning in one's life are entirely the product of individual choice grounded in subjectivity. But we do not create meaning *ex nihilo*. We construct it from the memories and other mental states we naturally have and need to navigate the world. Structures of meaning emerge from how we revise and use the cognitive and emotional content of our episodic memories in responding to the events that surround us.

What makes our actions authentic or inauthentic is how they affect our selves. Neither Taylor nor Elliott endorses an essentialist conception of the self. Still, the idea of authenticity as self-discovery suggests that the self is an intrinsic property that is unalterable and complete, and that an action is authentic when it reflects this property. Yet if our true selves are a work in progress, if we do not discover but construct and reconstruct them in response to external events, then authentic actions are those that contribute to this process. The self endures, not because it is immune to these events, but because of how we respond to them. Updating memories, including erasing them, can be part of this process when it enables us to respond to and imagine current and future events. Retaining a memory that caused one to ruminate on the past could impede this mental capacity and threaten authenticity. Erasing it could support this capacity as a core feature of the self.

Authenticity is primarily a property of prudential agency. It may not always track moral agency. There may be cases in which erasing an unpleasant memory of an action was consistent with one's interests and values but resulted in harm to others. Self-regarding reasons for erasing the memory may conflict with other-regarding reasons for retaining the memory to remind one of the harmful act. The moral reasons for retaining or erasing an unpleasant memory – as distinct from a traumatic memory – of an action depend less on how it affects one's own interests and more on how it affects the interests of others, specifically their interest in avoiding harm. Questions about the moral permissibility or impermissibility of erasing memories revolve around social behavior rather than authenticity. The concept of authenticity thus may be limited in assessing reasons for and against memory modification.

Even if modifying memories was consistent with one's interests and values, some might claim that taking a drug to produce this effect would undermine the person's authenticity. It is not just the outcome of memory modification that matters, but how one produces it. Something external, or alien, to the person would be the agent of change. According to Elliot, alienation consists in "an incongruity between the self and external structures of meaning – a lack of fit between the way you are and the way you are expected to be, say, or a mismatch between the way you are living a life and the structures of meaning that tell you how to live a life" (Elliott, 2000, p. 8). He expresses the same concern as the President's Council about memory modification. But if we create and identify with these structures of meaning after critical reflection, and if erasing a memory is consistent with these structures, then voluntarily taking a drug or undergoing a procedure to erase the memory would not be alien to the person. Deciding to erase it can be consistent with the motivational states one endorses as the genuine source of one's actions. It can be an authentic act. If a person decides that he would be better off without the memory, and that erasing it is consistent with his interests and values, then he would be the agent of change. The memory-erasing drug or procedure would only be the means through which he effects the change. The act would conform to his second-order desire that certain unpleasant memories not unduly influence the first-order mental states in his reasoning and decision-making.

More controversial is whether it would be ethically justifiable for one person to give proxy consent for a medical professional to erase another person's pathological memory. This would be a case where the memory significantly impaired the person's cognitive and emotional capacity to give informed consent for memory modification. Some would argue that any direct intervention in the brain to modify memory or other mental states without the person's consent should not be permitted because it would violate the person's cognitive liberty (Mill, 1859/1974, ch. 2; Bublitz and Merkel, 2014). But if the memory is the cause of psychic suffering and mental impairment undermining the capacity for informed consent, then it is questionable how much cognitive liberty the person retains. Removing the source of the mental impairment by erasing the memory could restore cognitive liberty rather than violate it. If the intervention were safe and effective, then reasons for permitting proxy consent for memory modification in cases of severe disorders of memory content could outweigh reasons for prohibiting it. It would be permissible when the memory caused maladaptive or pathological thought and behavior and significantly reduced the affected person's quality of life. This situation could occur in a case of severe PTSD caused by a memory of a traumatic experience.

Proxy consent for erasing a nontraumatic unpleasant memory would be significantly different and more difficult to defend. This may involve a patient with a psychiatric disorder and impaired decisional capacity with a memory of a past misdeed. The memory would not cause his disorder but could exacerbate his emotional symptoms. Proxy consent for erasing it would be objectionable for two reasons. First, it is doubtful that the memory would cause mental impairment disabling the capacity for reasoning and decision-making necessary for consent. Although the memory may adversely affect his quality of life, cognitive impairment motivating a request for proxy consent to erase a memory would be caused not by the memory but by other factors. The memory would not likely preclude his capacity for consent. Second, the risk of adverse effects from intervening in the brain to erase an unpleasant memory could outweigh the benefit. Even brain interventions considered safe and effective entail some risk.

Given the greater degree of mental impairment in a traumatic fear memory compared with an unpleasant memory, the probable benefit from erasing the second type of memory would be less than the probable benefit from erasing the first type. The permissibility of proxy consent for memory modification would depend on the safety of the intervention, the extent of mental impairment from the memory, the probability of positive and negative effects of the intervention, and evidence that it was in the person's best interests (Buchanan and Brock, 1990, pp. 112ff.). In principle, these conditions could be met in a case of traumatic memory, but not in a case of nontraumatic memory.

Conclusion

PTSD and other psychiatric disorders can be characterized as disorders of memory content. The emotionally charged representation of the trace causes hyperactivity in the fear memory system resulting in pathological and maladaptive thought and behavior. The problem is not an inability to form, store and retrieve memories but an inability to eliminate them from the brain and mind. I explained the mechanisms through which a fear memory in the basolateral amygdala might be erased, emphasizing that this intervention is hypothetical and that the challenges of localization and selectivity would have to be overcome to achieve this goal. I also explained how chronic pain could be characterized as a disorder of memory content. Like techniques that update fear memories with nonfearful information, behavioral techniques that updated information associated with pain memories and gradually replaced it with new information might weaken or remove the memory trace from the pain network. These interventions might relieve pain noninvasively by modulating neural and mental processes involved in conditioning and expectation. Although noninvasive cognitive behavioral therapy to treat chronic pain and other disorders of memory content would be safer than invasive therapies, it could leave open the possibility of stimuli reactivating the pain memory trace.

There are questions about how modifying memory would change a person's motivational states and actions. There are also questions about how these changes would influence judgments about whether memory modification could be ethically justified. Weakening or erasing a pathological fear memory could be justified because of the adverse effects on a person's mental life. Although the reasons would be weaker, erasing nonpathological unpleasant memories could also be justified if they interfered with practical reasoning and decision-making. This would be an authentic act if the person decided to erase the memory following critical reflection on the reasons for and against it, and the action was consistent with his interests and values. Erasing an unpleasant memory would be an irrational act if it meant acting against one's best interests. But provided that the means through which it was erased was safe, the act was voluntary, and any change in one's behavior did not harm other people, it would be permissible. Modifying memory would not be justified if it eroded the capacity to have and respond to moral emotions and resulted in a weakening or loss of moral sensibility in recognizing and responding to the rights, needs and interests of others.

Modifying memories may lie in the middle range of the continuum from therapy to enhancement. Whether we describe it as therapy or enhancement may depend on whether the memory is pathological, disturbing or just unpleasant. It may depend on the extent to which the emotional content of the memory interferes with normal thought and behavior. In one sense, erasing a pathological or disturbing memory can be described as therapy because it reduces psychological harm caused by the memory. It is a form of therapeutic forgetting. In another sense, erasing such a memory can be described as enhancement because it removes an obstacle to information processing and can improve rational and moral agency. Modification in disorders of memory content may be therapy when the intervention removes a harmful memory. Modification in disorders of memory capacity may be therapy when the intervention improves or restores memory function to normal levels. It may be enhancement when it raises these functions above these levels. How one describes these interventions depends on the type and extent of memory impairment and the intended outcomes. I discuss different interventions for disorders of memory capacity in Chapter 5.

5   Disorders of Memory Capacity and Interventions

Most disorders of memory are disorders of memory capacity. They involve impaired capacity to encode, consolidate, retrieve or reconsolidate information about events and facts. Anterograde amnesia is the inability to encode and consolidate new memories. Retrograde amnesia is the inability to retrieve or reconsolidate stored memories. These conditions may result from traumatic brain injury or cerebrovascular accidents causing ischemia, hemorrhage or dysfunction in the hippocampal complex and cerebral cortex. Impaired working and spatial memory may result from anterograde or retrograde amnesia because they rely on stores of episodic and semantic information in these brain regions. Amnesia may also be psychogenic and triggered by a person's response to psychosocial stress. Transient global amnesia and dissociative amnesias are temporary and typically resolve in complete recovery. But they can leave a gap in one's mental states between the onset and resolution of the episode.

Damage to the striatum or cerebellum can impair procedural memory. Alzheimer's often begins with degeneration in the hippocampal-entorhinal circuit and disruption of working and spatial memory. This neurological disease involves gradual degeneration in cortical and sub-cortical regions and eventually results in the loss of all forms of declarative and nondeclarative memory.

In this chapter, I focus on chronic disorders of memory capacity and interventions to treat them. I address differences in the use of "therapy" and "enhancement" in memory modification. In most cases, brain interventions designed to treat memory disorders can be described as therapies. A smaller number of interventions designed to improve normal memory function can be described as enhancements. I discuss how certain drugs and transcranial stimulation may improve or raise memory above a normal level. Drugs that increase circulating levels of neurotransmitters to enhance memory may involve risks, and these need to be weighed against any positive changes in memory function. I also consider deep brain stimulation as an experimental treatment for memory impairment in early-stage Alzheimer's. High-frequency deep brain stimulation of circuits consisting of the hippocampus, fornix and entorhinal cortex could activate synaptic connectivity and LTP to improve spatial and working memory. But this is an invasive procedure that could cause bleeding, infection and neurological and psychological sequelae.

I explain why devices external to the brain could not replace it as the source of memory. Then I explore the potential of a hippocampal neural prosthetic to improve memory encoding and retrieval for people with damaged hippocampi. I consider the possibility of a prosthetic replacing the hippocampus, how it would restore memory function and how this or similar devices could decode the cognitive and emotional content of episodic and emotional memories. Having an artificial memory system implanted in the brain could have far-reaching psychological and social implications.

Because of differences in their mechanisms of action, cognitive and behavioral therapies may have more therapeutic potential for disorders of memory content than for disorders of memory capacity. Brain implants may have more therapeutic potential for the second type than for the first. Despite the differences between the two types of memory disorders, all interventions designed to treat them involve modifying information in the brain constituting the memory trace. Some of these interventions aim to disrupt or remove information from the brain and mind. Others aim to make information processing more efficient or restore it to a normal level. The common goal is to regulate dysregulated memory systems and modify memory to restore and maintain its adaptive purpose.

Restoring or Improving Memory: Therapy or Enhancement?

In Chapter 4, I claimed that erasing a disturbing memory could be a form of therapeutic forgetting. I also claimed that it could be a form of enhancement because removing the memory could improve information processing and decision-making. In addition, I claimed that improving or restoring memory to a normal level was therapy, and that raising it above this level was enhancement. Still, there are questions about whether memory modification should be described as therapy or enhancement.

Some forms of memory modification may be difficult to categorize because they fall within the broad middle range of memory function between exceptional recall and profound amnesia. Intuitively, the more impaired memory capacity is, the more plausible it is to describe an intervention to improve it as therapy. The less impaired the capacity is, the more plausible it is to describe the intervention as enhancement. Interventions that improve normal memory function are clearly enhancements. Interventions that improve memory capacity in cases of severe impairment are clearly therapies. This distinction may be more difficult to draw as people age and experience increasing degrees of cognitive impairment associated with declining working memory and memory retrieval. What matters in medically and ethically assessing brain interventions for disorders of memory capacity is not whether one describes them as therapies or enhancements. Rather, what matters is the degree to which they improve memory and promote a person's functional independence.

"Memory enhancement" appears in the title of a paper by Nanthia Suthana and coinvestigators reporting on a study testing the effects of deep brain stimulation on some patients with epilepsy (Suthana, Haneef, Stern et al., 2012). The study showed that electrical stimulation of the entorhinal cortex improved working and spatial memory. If the study participants were impaired in these two types of memory, and the results improved memory function to approximately normal levels, then it would be more accurate to describe this application of DBS as experimental therapy rather than enhancement. The investigators may have considered any improvement as "enhancement," regardless of baseline memory function. Researchers in a more recent study involving stimulation of the temporal cortex of patients with no memory impairment state that the technique "improves" memory encoding and retrieval (Ezzyat, Wanda, Levy et al., 2018). The different descriptions by the investigators in these two studies suggest that an improvement in memory function could be described as enhancement or therapy. In a very different study, researchers showed that methylphenidate and modafinil improved the performance of accomplished chess players when they were not under time constraints. The fact that these subjects already had a high level of working memory indicates that raising this level to produce more effective decision-making was an enhancement (Franke, Gransmark, Agricola et al., 2017).

"Cognitive enhancement" generally refers to direct or indirect interventions in the brain that improve attention, concentration and information processing above normal levels (Jotterand and Dubljevic, 2016). These in turn improve executive functions such as reasoning and decision-making. These functions primarily involve working memory. They also involve episodic and semantic memory because working memory depends on these other types of declarative memory. Enhancing memory function is a form of enhancing cognitive function.

There are three main concepts of cognitive enhancement: augmenting, diminishing and optimizing. The first concept considers brain interventions as enhancements when they improve a cognitive function by increasing its ability to do what it normally does (Harris, 2007; Bostrom and Sandberg, 2009; Savulescu, Sandberg and Kahane, 2011). An enhancement is an "intervention designed to improve human form or function beyond what is necessary to restore or sustain good human health" (Juengst, 1998, p. 29). The second concept of enhancement claims that some functions can be improved by diminishing the extent of what they do and their effects. Erasing an unpleasant memory to improve memory processing and decision-making is an example of this type of enhancement (Earp, Sandberg, Kahane et al., 2014). The third concept of enhancement refers to an intervention that "aims at optimizing a specific class of information processing functions: cognitive functions, physically realized by the human brain" (Metzinger and Hildt, 2011, p. 245).

A broad optimizing concept is probably most consistent with people's intuitions about cognitive enhancement. The goal of modifying memory and other cognitive functions is not just to improve performance on certain tasks but more generally to improve the role of memory in thought and behavior. This is more likely to occur when neural processing is neither underactive nor overactive and the brain and mind are not overloaded with too much information. Enhancing memory does not simply mean increasing levels of neurotransmitters or other factors associated with it. Depending on the amount of information in the brain at a given time, optimal levels of memory could be produced by augmenting or diminishing this information. "Optimal" suggests that there are limits to the extent to which we can improve these functions by increasing circulating levels of neurotransmitters or activating neural circuits (Agar, 2014). By overly manipulating these functions, "healthy individuals run the risk of pushing themselves beyond optimal levels with hyperdopaminergic and hypernoradrenergic states, thus vitiating the very behaviors they are striving to improve" (Urban and Gao, 2014, p. 1).

My discussion of hyperthymesia in Chapter 2 and how it can impair agency by impairing working and prospective memory underscores the importance of optimal levels of episodic and semantic memory. An excess of these types of memory can overload the brain and mind with representations of events and facts. It can interfere with efficient processing of this information and effective decision-making. One hypothesis for balancing demand and capacity in working memory is that the brain may be able to process more information as the demands of the brain and organism increase (Klingberg, 2008). But the information must be relevant to these demands, which was not the case in the hyperthymesia of Solomon Shereshevskii, Jill Price, or Borges's fictional character Funes. An increase in the level of information would have to occur incrementally to allow the brain and mind to adjust to this change. How well they adjust would also depend on the nature of the demands and the type of information relevant to them. Some demands may require more, or less, of one type of memory than others. Interventions designed to treat disorders of memory capacity must ensure optimal levels of encoding, consolidation, storage, retrieval and reconsolidation to achieve their therapeutic goal.

Psychopharmacological Interventions

Drugs aimed at improving the capacity to form, store and retrieve short- and long-term memories target certain neurotransmitters associated with LTP and protein synthesis. Many of these drugs aim to inhibit the breakdown of acetylcholine or to stimulate glutamate in the brain. Some have been tested in clinical trials for Alzheimer's disease. They include the acetylcholinesterase inhibitor donepezil and the glutaminergic NMDA receptor antagonist memantine. Except for some modest benefit in some patients with early symptoms, pharmacological interventions to restore memory function in this and other neurodegenerative disorders have been a failure. Despite advances in neuroimaging identifying the source of dysfunction and designing drugs to target it, there is no evidence that they can delay, stop or reverse dementia (Matthews, 2011, p. 116). They may also have serious side effects. For example, increasing levels of glutamate in the brain may cause seizures (Matthews, 2011, p. 116). Protein kinase modulators for Alzheimer's could also be toxic above a certain level. This limits treatment to low doses that may not be effective in ameliorating impaired memory and other cognitive symptoms of the disease.

The critical question is whether drugs can modulate the relevant neurotransmitters and excitatory and inhibitory mechanisms associated with memory. This could ensure a balance between learning and forgetting through balanced memory formation, storage and retrieval. To date, no psychotropic drug has been able to do this in humans to any significant degree. Nor is there any evidence that these classes of drugs could restore memory functions lost from amnesia. The cure for the semantic amnesia of the inhabitants of Garcia Marquez's Macondo pulled them out of the "quicksand of forgetfulness." But no pharmacological cure for amnesia and other disorders of memory capacity is forthcoming in the foreseeable future. Still, it is worth considering different classes of drugs that might have therapeutic potential.

A drug that strengthened synaptic connectivity and LTP by increasing circulating levels of PKM-zeta might improve the ability to encode, consolidate and retrieve memories. It could have an activating effect on this protein, in contrast to the blocking effect of ZIP on unwanted memories discussed in Chapter 4. Ampakines can increase LTP by activating glutamate through the alpha-amino-3-hydroxy-5-methyl-4-isoxazolepropionic (AMPA) receptor (Lynch and Gall, 2006). Dosing would be critical, though, because increased activity in glutaminergic pathways could cause seizures or other adverse neurological events from overly excitatory neurons.

Other drugs might be able to strengthen synaptic connectivity and increase the ability to form new memories and retrieve existing memories by increasing levels of CREB in the brain. They would have to be specific enough to target only the memory-activating and not the memory-blocking version of CREB (Tully, Bourtchouladze, Scott et al., 2003). These drugs might prevent hyperthymesia because this and other transcription factors allow only a certain amount of information in the brain. Drugs modulating memory-activating and memory-blocking CREB could ensure a balanced level of memory formation and retrieval between deficit and excess. Studies of the ampakine farampator have been significant because they have been conducted using human rather than animal models. The results of a study involving healthy elderly volunteers showed that farampator improved working memory but impaired episodic memory (Wezenberg, Verkes, Ruigt et al., 2007). This raises the question of whether chronic use of this drug would result in limited rather than improved working memory because of its dependence on episodic and semantic memory.

Additional human trials are necessary to produce solid data on the short- and long-term effects of these drugs on different types of memory and stages of the memory process. The results of the farampator study suggest that there may be trade-offs in pharmacological modification of memory. A drug may improve the capacity to form new memories but not improve or even interfere with the capacity to retrieve memories. It may ameliorate anterograde amnesia but not retrograde amnesia. It is not known whether a drug would improve consolidation, retrieval and reconsolidation to the same degree. LTP and protein synthesis may have different effects on these stages. While it is necessary for reconsolidation, protein synthesis may not be necessary for consolidation (Gold, 2008). Drug design may be limited to mechanisms in one but not other stages of memory. Nor is it known whether memory-modifying drugs would have different effects on episodic, semantic or working memory in people with amnesia from neurological disorders versus those with traumatic brain injury (Izquierdo, Cammarota, Medina et al., 2004). To be effective in improving or restoring memory, drugs would also have to be specific enough to distinguish emotionally charged from non–emotionally charged memories in activating the neural mechanisms underlying them. Specificity may be the most difficult obstacle for memory-modifying drugs to overcome in determining their safety and efficacy.

Although I have defended upholding a general distinction between therapy and enhancement, some drugs used to enhance cognitive functions may yield clues to how they can improve memory. Stimulants such as methylphenidate can increase circulating levels of dopamine and improve working memory. Other drugs targeting dopamine and other neurotransmitters may have broader effects on different memory types and stages of the memory process. "Activation of the dopamine D1 receptor system enhances synaptic plasticity in the hippocampus" (Matthews, 2011, p. 116). This could improve memory formation and retrieval by improving communication between hippocampal and cortical networks. In addition, "blocking some actions of the neurotransmitter histamine, which helps to regulate the release of monoamines [including dopamine] through its H3 receptor, can facilitate measures of memory in animal models and, in particular, reverse the partial amnesia induced by scopolamine, an antagonist of the neurotransmitter acetylcholine, which acts at the muscarinic cholinergic receptor" (Matthews, 2011, p. 116).

Drugs that increased levels of some neurotransmitters while reducing others might improve the capacity for memory formation and retrieval. Like glutamate, chronic high circulating levels of dopamine in the brain could be toxic and cause seizures or other adverse neurological and psychological events. Modafinil is a drug initially designed for narcolepsy. It has been used to enhance cognitive capacities by increasing alertness and attention despite sleep deprivation. Its enhancing effects are due to increased levels of dopamine in the striatum, nucleus accumbens and other brain regions. Raising levels of this neurotransmitter could improve working memory. But it could also activate the brain's reward system and make one susceptible to addictive behaviors such as gambling and hypersexuality (Volkow, Fowler, Logan et al., 2009). Depending on their side effects and the extent of amnesia or other impaired memory capacities, drugs targeting neurotransmitters associated with learning and applying information may be limited in improving memory. Their efficacy depends on maintaining acetylcholine, dopamine, glutamate or other neurotransmitters within optimal levels.

Future studies may show that existing or emerging psychotropic drugs could be selective enough to target specific memory systems and stages to treat amnesia and other forms of memory impairment. It is possible that they could restore and maintain optimal levels of excitatory and inhibitory neural mechanisms and strike the right balance of learning and forgetting.

The lack of solid evidence that these drugs can significantly improve memory function and avoid toxic and addictive effects points to the need for alternative interventions as therapies for disorders of memory capacity.

Neurostimulation and Memory

Increasing activity in brain circuits rather than neurotransmitters may have greater potential to improve or restore memory functions to normal levels. TMS pulses may induce LTP in the targeted neurons and synapses and promote neuroplasticity necessary for memory formation (Nitsche, Boggio, Fregni et al., 2009). As noted in Chapter 4, however, the effects of TMS may be limited because the pulses penetrate only the first centimeter of cortex. This raises questions about the extent to which this technique could activate memory circuits. It may activate the prefrontal cortex and improve working memory. It may also activate other regions of the neocortex and improve retrieval of semantic memory. But if the pulses could not penetrate deeper structures in the hippocampal complex, then they might not affect consolidation or retrieval of episodic memory. Each of these memory stages depends on the hippocampus and other structures in the medial temporal lobes. TMS would be safer than DBS because it is less invasive and avoids the risks of intracranial surgery and implants in brain tissue. But its therapeutic effects would probably be limited to working memory and retrieval mechanisms of semantic memory.

Like the drug farampator mentioned earlier, TMS may involve trade-offs in its effects on semantic memory. Researchers have used a variant of TMS, transcranial electrical stimulation (tES), to test the learning and application of mathematical information. They found that stimulating an area of the subjects' prefrontal cortex impaired learning new information but enhanced the application of what they learned. Stimulating an area of the parietal cortex had the opposite effect of enhancing learning while impairing the ability to apply the new information (Iuculano and Cohen Kadosh, 2013). The upshot of this and similar studies is that improving some cognitive functions through psychotropic drugs or electrical stimulation may impair other functions. This pertains to memory as well. It supports the established view that memory systems may be dissociable or interdependent. Brain interventions aimed at improving one type or stage of memory may not have the same effect on and may impair others. The studies also support the position that there are optimal levels of memory formation, storage and retrieval. Whether memory modification could restore and maintain balanced learning and forgetting would depend on the extent of impairment and how the intervention affected the critical neurobiological mechanisms.

DBS can influence a broader range of memory functions than TMS. It can penetrate deep structures of the brain involved in processing information about the past and using this information to simulate future events. In Chapter 4, I mentioned a study showing how deep brain stimulation of structures in the hippocampal complex could disrupt episodic memory retrieval and reconsolidation (Merkow, Burke, Ramayya et al., 2017). DBS can also have the opposite effect of improving memory impaired by brain injury or neurological diseases like epilepsy and dementia. Differences in its weakening and strengthening effects on episodic memory can be explained by differences in the frequency of the electrical current and the level of metabolic activity in its neural targets. Erasing memory would involve inhibiting or downregulating metabolically hyperactive nuclei. Improving memory in approximating a normal level of information processing would involve activating or upregulating metabolically hypoactive nuclei. The stimulation parameters would be set to influence CREB, LTP and protein synthesis in producing the different effects. Functional imaging before and after the intervention could confirm changes in neural activity and synaptic connectivity associated with the memory trace.

Like memory erasure, memory enhancement would involve the challenge of specificity in targeting certain brain regions without affecting others. Improving memory may be more difficult because disorders of memory capacity typically involve interaction between distributed neural networks. DBS is more invasive and entails more risk than other forms of memory modification. Again, there is the risk of hemorrhage from the intracranial surgery to implant the electrodes. In addition, there is the risk of infection and expanding adverse effects on normal neural circuits and tissue in adjacent brain regions. Depending on the circuits that are stimulated, there is also the risk of inducing hypomania and other pathological behaviors from overactivating the reward system. But these risks need to be weighed against the potential benefit of improved memory from neurostimulation (Castrioto, Lhommee, Moro et al., 2013; Lozano and Lipsman, 2013).

The hippocampal-entorhinal circuit has been the main target of DBS as an experimental treatment for disorders of memory capacity. The study by Suthana and coauthors showed that DBS of the entorhinal area can improve learning and spatial memory in patients with temporal lobe epilepsy. This effect has been replicated in other studies (Fell, Staresina, DoLam et al., 2013). In an earlier clinical trial, stimulation of the fornix increased glucose metabolism in this region and slowed cognitive decline in some of the six participants with early-stage Alzheimer's disease (Laxton, Tang-Wai, McAndrews et al., 2010). A more recent study involving Alzheimer's patients over the age of sixty-five showed that participants who received stimulation of the fornix experienced less cognitive decline over a year than those who did not receive it (Lozano, Fosdick, Chakravarty et al., 2016). Improvement in spatial and working memory in some participants in these studies has enabled them to maintain some level of social and professional activity despite memory impairment associated with the disease.

A different group of researchers stimulated a different region of the brain, the lateral temporal cortex (LTC), to enhance memory in subjects being monitored for epilepsy. The subjects in this study did not have memory impairment. Unlike the more standard open-loop DBS, the frequency of stimulation in closed-loop DBS is not constant and fixed but responds in real-time to changes in the brain. The researchers used this technique in this brain region to identify neural activity associated with poor encoding and retrieval of words. They showed that stimulating the LTC could predict successful recall of the word. This is an example of how electrical stimulation of the brain can improve semantic and possibly other types of memory (Ezzyat, Wanda, Levy et al., 2018). By showing how DBS can enhance encoding and retrieval of semantic information, the study is particularly significant because it shows the potential of DBS to improve function at different stages of the memory process. Another study involving human subjects demonstrated that direct brief stimulation of the amygdala improved declarative memory for specific images without eliciting an emotional response (Inman, Manns, Bijanki et al., 2018). Less invasive electrical stimulation of temporal and frontal-temporal cortices improved working memory in older adults (Reinhart and Nguyen, 2019).

In Chapter 4, I mentioned a study in which researchers used optogenetics to erase memory in mice. Optogenetics could also be used to improve or restore memory. This would involve manipulating opsins to activate rather than inhibit neural circuits, LTP and protein synthesis. In one study, optogenetic activation of memory engram cells enhanced synaptic connectivity and memory retrieval. The researchers produced this effect following protein synthesis inhibitor-induced retrograde amnesia (Ryan, Roy, Pignatelli et al., 2015). The research in this area of memory modulation is much less developed than it is for DBS and has been conducted only in animal models. I focus on DBS in exploring some of the ethical issues in attempts to improve or restore memory.

Although the initial results of using DBS for impaired memory have been promising, data from studies involving a small number of patients are far from conclusive. Phase II and Phase III clinical trials with a larger number of research subjects are necessary to know whether its therapeutic potential can be realized. There is no evidence that this technique can stop or reverse the progression of neuropsychiatric disorders. Nevertheless, some studies suggest that it could have neurogenerative effects and improve or restore memory by strengthening synaptic connectivity.

Results of a study of DBS for early motor symptoms in Parkinson's disease suggested that stimulation of the subthalamic nucleus at an earlier stage of the disease was superior to pharmacotherapy in relieving symptoms (Schuepbach, Rau, Knudsen et al., 2013). The study also suggested that stimulation could induce neuroplasticity and possibly slow progression of the disease. Similarly, continuous electrical stimulation of the hippocampal-entorhinal circuit might induce endogenous repair and growth mechanisms in this region of the brains of patients with Alzheimer's. Activation of nuclei in this circuit could activate LTP, protein synthesis and synaptic connectivity regulating different stages of memory. This is significant because the hippocampal-entorhinal circuit is one of the first brain regions affected by tau pathology and beta-amyloid neurofibrillary tangles associated with cognitive impairment in this disease (Bejanin, Schonhaut, La Joie et al., 2017; Aldehri, Temel, Alnaami et al., 2018).

By activating repair and growth mechanisms in the hippocampal-entorhinal circuit, it is possible that early application of DBS could improve or restore memory and prevent further memory loss in AD. It would do this by strengthening and maintaining synaptic connectivity, LTP and other neural mechanisms enabling episodic, semantic, working and spatial memory. Neurostimulation could restore retrieval mechanisms allowing a patient to access memories still in storage. In aging or diseased brains, the technique could enable retrieval of memories that were "lost" because they could not be retrieved. The general effect on these mechanisms would translate into specific positive effects on different types and stages of memory.

Ideally, the outcome of this intervention would avoid any trade-offs between learning new information and applying it. It would not strengthen consolidation at the cost of impairing retrieval, or vice versa. Instead, it would strengthen both processes and possibly reconsolidation as well. Reversing brain damage and neurodegeneration is not likely in the foreseeable future. DBS can improve some memory functions but not alter the underlying pathophysiology of memory disorders. Still, it is instructive to explore some of the possible neurological and psychological effects of continuous stimulation of neural circuits mediating memory.

A person with early-stage Alzheimer's participating in a DBS trial involving stimulation of the hippocampal-entorhinal or hippocampal-fornix circuit may experience improvement in working and spatial memory. It may increase her neural and mental ability to encode and learn new information. However, if the technique does not reverse or stop the disease process, then the improvement will result entirely from electrical stimulation of the targeted circuits. As the disease progresses, synaptic connectivity will weaken further, and any positive effects of DBS on memory networks will diminish. Competent patients who experience improvement in their memory from neurostimulation may overlook or ignore the fact that the positive change in memory is only temporary and that they will eventually lose it. Even when researchers make patients aware of the limitations of this experimental treatment when they consent to undergo it, hope and other emotions may cause patients to have unreasonable expectations about the duration and extent of the improvement. Some may form the erroneous belief that their improved memory will continue indefinitely. This belief could make them susceptible to harm when the positive effects diminish and then end. The course of the disease could thwart their desire and hope for sustained memory function.

The change from a state of impaired memory to a state of improved memory, and then back again, could make it difficult for the patient to adjust to the change. It could cause her to suffer twice: before the neurostimulation and after its effects had ceased. The second period of suffering could be worse than the first because the progressive loss of memory would follow a period of improvement. It would defeat any positive expectation the patient might have developed from the therapeutic effect of the stimulation. The psychological adjustment could be especially difficult for Alzheimer's patients who may have undergone personality changes and become irritable and anxious. While there are clear and convincing reasons to improve and sustain memory and functional independence, awareness that the improvement was not permanent could limit its benefit for the patient. If the improved memory made one more aware of one's disease, then it could cause one to dwell on the inevitable bad outcome. The harm caused by the disease and a patient's awareness of its progression could outweigh the temporary benefit.

The pathophysiology of memory impairment caused by neurological diseases may be different from that of memory impairment caused by traumatic brain injury. It is not known whether patients in one group would experience more improvement from neurostimulation of memory circuits than those in the other group. If DBS cannot activate repair and growth mechanisms to the point of restoring brain function, then the memory deficits in these two conditions may be equally irreversible. Brain damage eventually results in neurodegeneration. Yet there may be a period in which brains with some but not extensive damage from trauma and not yet showing signs of degeneration may respond more readily to neurostimulation than brains with Alzheimer's. Nevertheless, DBS does not replace memory circuits but only compensates for dysfunction in them. The technique would probably not have any positive effects in circuits that were completely dysfunctional. Patients with extensive brain damage would likely not recover memory functions from neurostimulation. The damage to Clive Wearing's brain from the herpes viral encephalitis may have been too extensive for neurostimulation to restore any of his declarative memory.

Consider now the hypothetical situation where deep brain stimulation could activate neural circuits and restore some degree of memory function in more advanced dementia. A surrogate could give proxy consent for the intervention. The extent of neurodegeneration at an advanced stage of the disease could make any potential memory improvement unlikely. Theoretically, some neural circuits mediating episodic memory may be intact. High-frequency stimulation might increase glucose metabolism and reactivate LTP and synaptic connectivity in these circuits. A patient could go from being severely memory impaired to having more memory. This might also increase his level of awareness and capacity to anticipate the future. Unless the effects could be sustained indefinitely, temporary memory improvement could be worse and more harmful for the person than if he had remained memory-impaired and unaware of his condition. It could make him aware that, once the positive effects of the stimulation stopped, he would return to his impaired state and progress to oblivion.

The two examples I have presented illustrate how improving episodic memory through neurostimulation may be both beneficial and harmful for patients at different stages of dementia. There are no therapies that can control the neurodegeneration causing this disorder. If neurostimulation enabled demented patients to recover even a limited degree of their capacity to anticipate the future, then this could cause them to suffer from the negative expectation of what is to come. Despite this possibility, there are compelling medical and ethical reasons for neurostimulation as an experimental intervention for early-stage dementia. It can improve some types of memory and enable functional independence for people affected by it, even if for only a limited period. There may be ethical reasons against neurostimulation for more advanced dementia. The probability of the technique increasing memory capacity at a later stage of the disease would be low. If it had this effect, then the potential psychological harm could outweigh the potential benefit. Unless DBS or other techniques could stop or reverse the pathophysiology of disorders of memory capacity, temporary memory improvement could have more of a net disvalue than value. The most favorable benefit–risk ratio for memory modification with DBS would be at an early disease stage where some neural circuits were still functional and responsive to stimulation.

In treating any disorder of memory capacity, imprecise stimulation or frequency above an optimal level might inadvertently generate false memories. The technique might activate circuits in a way that would

distort information in them. It might result in neural and mental representations that did not correspond to actual events in the patient's life.
Wilder Penfield's probing of the temporal cortex elicited recall of episodic memories in some of his epilepsy patients. But inducing recall
could have an adverse psychological impact on patients with early-stage
dementia who are emotionally labile. Although they have lost a significant degree of their episodic and semantic memory, they may retain
enough mental capacity to suffer from confusion about and emotionally
charged reactions to these mental representations. The possibility of
neurostimulation triggering false memories and negative psychological
responses to them should be factored into calculating potential benefit
and risk in attempting to improve memory through this technique.

Unauthorized third parties could induce false memories by disrupting
the radiofrequency of the electrodes implanted in the hippocampal-
entorhinal or hippocampal-fornix circuit. Hackers violating neural implant
security could create neural and mental representations of past events that
never occurred. This and other effects could impair cognition and mood
in people with the implant in their brains (Pycroft, Boccard, Owen et al.,
2016). Hackers could deliberately distort the processing of information
about the patient's past and alter the content of true memories as well.
Besides disrupting encoding mechanisms, they could disrupt retrieval
mechanisms, causing them to be underactive and exacerbating retrograde
amnesia. Or they could cause them to be overactive and induce a type of
hyperthymesia in uncontrolled recall. Brainjacking could alter the content
of memories by disrupting normal communication between the hippocampus and amygdala. This could generate negative emotional valence in
memories, which in turn could induce stress and fear responses resulting
in further psychological harm to patients. These outcomes would be in
addition to the potential to disrupt stimulation in producing pathological
behaviors or reinforcing existing ones.

This type of interference would require a high level of technical
sophistication. But a determined hacker might be able to do it. Ensuring
implant security by making the electrodes and pulse generator immune
to this interference would be a challenge for neurostimulation device-
makers over and above ensuring that the device could safely and effectively modulate memory circuits. Despite measures that could be taken to
reduce this risk, it could not be eliminated.

It is not known how long continuous deep brain stimulation could
sustain a degree of memory capacity sufficient for a person with a disorder
to remain functionally independent. This will depend on how much neural
circuitry is preserved and the effects of activating them on different types of
memory. Given that brain function decreases as the degenerative effects of

brain injury and dementia increase, and that the effects of DBS depend on a certain degree of brain function, any improvement the technique could produce might be limited. If the circuits have degenerated to the point where they have little or no metabolic activity and are completely dysfunctional, then it is unlikely that deep brain stimulation could restore or improve memory to any degree.

External Devices and the Extended Mind

Neurostimulation involving brain implants can compensate to some degree for memory dysfunction and improve memory capacity. DBS and other memory-modifying techniques can supplement but not supplant or replace neural circuits mediating memory. There must be some preserved function in these circuits for neurostimulation to have any effect on them. DBS cannot encode, consolidate or retrieve memories on its own. Some might claim that external devices such as advanced smartphones could substitute for the brain in processing information constituting memory. Presumably, this information could be realized just as well in an external device as it is in the brain. When programmed in the right way, the device could execute all functions in the memory process.

Entering information in and retrieving it from a device requires a certain level of cognitive capacity. This capacity rests not on the device but on normally functioning neural networks. Nor can any object external to the brain substitute for these networks as the foundation of the meaning a person attributes to memory. This meaning is the product of episodic and emotional memories contextualized by the person's relation to the environment. It is also the product of her autonoetic consciousness and experience of mental time travel. No device can replicate these subjective aspects of memory. These points highlight the limitations of artificial memory systems and undermine the claim that an external device could replace a natural brain in generating and sustaining memory.

The idea that a device could replace the brain as a repository of memory is related to the extended mind theory proposed and defended by Andy Clark and David Chalmers (Clark and Chalmers, 1998; Rowlands, 2010, pp. 58–67). They adopt an externalist conception of the mind in which objects within the environment function as part of it. Because external objects influence cognition, the mind and the environment operate as a coupled and complete cognitive system. The mind extends beyond the confines of the skull into the external world. Clark and Chalmers present a thought experiment to illustrate and support the extended mind theory. Fictional characters Inga and Otto are traveling separately and simultaneously to a museum. Inga retrieves spatial memory internally processed by her brain for mental directions. Otto has a belief about the location of the museum. But because he has Alzheimer's disease and some memory impairment, he has entered information about the museum in a notebook and consults it to confirm how to get there. Retrieving spatial memory from the notebook will enable him to arrive at his destination. It operates like an external GPS that compensates for the information failure in the internal hippocampal-entorhinal circuit of his brain. This thought experiment suggests that memory need not be realized in the brain but could be realized in an object external to the brain. It suggests that the notebook could replace the brain as the source of Otto's memory and function well enough for him to access the information he needs to get to the museum.

Among critics of the extended mind theory, Fred Adams and Ken Aizawa argue that extended mind theorists commit the coupling-constitution fallacy by inappropriately making "an object cognitive when it is connected to a cognitive agent" (Adams and Aizawa, 2010, p. 67). Clark and Chalmers commit this fallacy by coupling Otto's memory with his notebook, then inferring that the notebook constitutes part of his memory system, Adams and Aizawa point out that "coupling relations are distinct from constitutive relations, and the fact that an object or process X is coupled to object or process Y does not entail that X is part of Y" (Adams and Aizawa, 2010, p. 67). The extended mind theory is problematic because it does not explain what cognition causally depends on or what constitutes cognition.

This fallacy is pertinent to the idea that a notebook or any device external to the brain could substitute for it in providing one with the information necessary to navigate the environment. The fact that the notebook is coupled with Otto's spatial, working, episodic, semantic and prospective memory systems does not entail that it constitutes part of these systems. Nor does it entail that these systems could be realized in it. It is not his notebook but his brain that provides him with the cognitive capacity necessary to enter information in the notebook and access information from it. As his Alzheimer's progresses, he will lose this capacity. At an advanced stage of the disease, the notebook will no longer serve as an external supplement to his dysfunctional internal memory system.

The extended mind hypothesis fails when philosophers try to apply it to memory. No object external to the brain can substitute for it in enabling declarative memory functions because no such object has the properties necessary for these functions. As the discussion of DBS to improve memory shows, internal objects such as brain implants can supplement but not supplant a natural brain. The salutary effects of these devices are limited by the level of preserved function in brain regions mediating memory. External devices cannot increase glucose metabolism and strengthen the synaptic connectivity necessary for all types of memory. These objects are at best a temporary informational aid compensating for declining function in memory-mediating neural networks. Without a certain functional level in these networks, the notebook is worthless as a memory aid. It cannot substitute for the brain in enabling one to process, store and retrieve information about oneself and one's environment. Neither the notebook nor any artificial information processing system by itself can retrieve the spatial information Otto needs to get to the museum. He can retrieve this information only insofar as his brain and cognitive capacity enable him to retrieve it. Beyond a certain point of cognitive decline, Otto or any actual Alzheimer's patient will no longer be able to consult or access information from the notebook or device. The neural networks necessary for him to perform these mental acts will have become completely dysfunctional.

The idea that Otto's notebook or another external device could replace the brain as the source of his memory is symptomatic of a failure to acknowledge the neurobiological basis of memory. The ability to manipulate information in electronic devices depends on episodic, semantic, working and prospective memory. These types of memory depend on normal functions in different regions of the brain. A notebook or device is only as smart as the brain allows it to be. The mind extends beyond the skull only insofar as the brain enables it. The mind is extended in the sense that the content of our mental states represents states of affairs external to us. But there is no object or system external to the brain that could replace it in generating and sustaining memory.

Hippocampal Neural Prosthetics

Novel brain implants other than those involved in DBS have the potential to be more effective in improving the capacity to form and retrieve episodic memories. These devices may provide a therapeutic alternative to electrical stimulation for anterograde and retrograde amnesia. They would do this through more precise identification and modulation of neural firing patterns in the hippocampal complex when it has been damaged by brain injury or disease. Current research indicates that devices recording and responding to these patterns can improve memory in this region when it is functionally intact or mildly impaired. Future research may enable devices to completely encode and retrieve episodic memories on their own without a natural hippocampal complex. Although it is still hypothetical, research may develop to the point where a device could not just improve function of innate neural circuits mediating memory but replace them when they have become dysfunctional from severe injury or advanced disease.

The most therapeutically promising of these devices is the hippocampal neural prosthetic. Funded by the US Defense Advanced Research Projects Agency (DARPA) and developed by research teams at the University of Southern California and Wake Forest University, the device consists of a multisite electrode array implanted in the area encompassing the hippocampal-entorhinal circuit. The array is linked to a very-large-scale integration (VLSI) biomimetic model that provides the necessary inputs and outputs to encode short-term episodic and working memory (Berger, Hampson, Song et al., 2011; Hampson, Song, Opris et al., 2013) (see Figure 5.1). For people with hippocampal

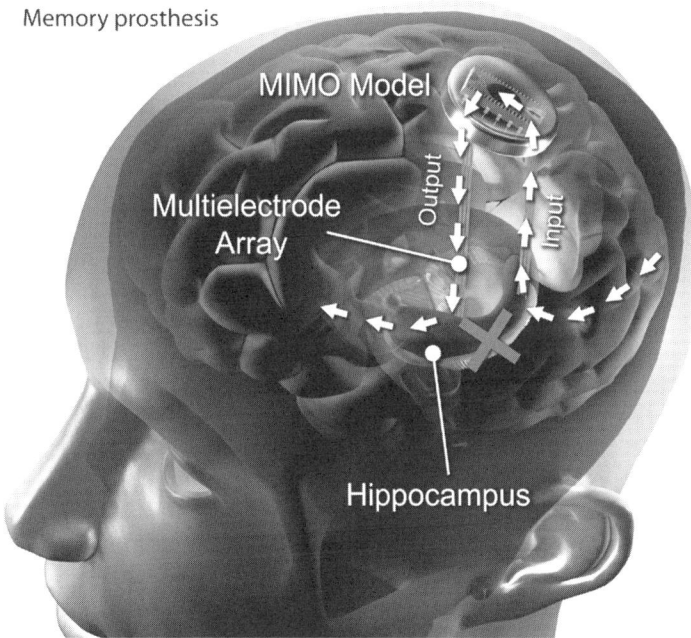

Figure 5.1 Conceptual representation of a hippocampal prosthetic for humans. The biomimetic multi-input and multi-output (MIMO) model implemented in a very-large-scale integration (VLSI) device performs the same nonlinear transformation as a damaged portion of the hippocampus ("X"). The VLSI device is connected "upstream" and "downstream" from the damaged hippocampal area through multisite electrode arrays. (with permission from Theodore Berger, image creator)

dysfunction caused by Alzheimer's disease or brain injury who do not respond to neurostimulation, a hippocampal prosthetic may allow some improvement in the capacity to learn new information and form new memories.

While earlier versions of this artificial system were tested only in animal models, researchers led by Robert Hampson tested the most recent version in humans in a proof-of-concept study. They enrolled epilepsy patients who were undergoing a mapping procedure to identify the source of their seizures. Based on a multi-input, multi-output (MIMO) nonlinear mathematical model, the researchers used the electronic prosthetic system to influence firing patterns of neurons in the hippocampus and improve memory encoding and recall (Hampson, Song, Robinson et al., 2018). Specifically, the team used this system to facilitate the encoding of short-term episodic memory and its use in working memory. By identifying neural firing patterns in the hippocampus indicating correct memory encoding and distinguishing them from patterns indicating incorrect encoding, the team used the prosthetic to strengthen the correct patterns and assist the brain in forming new memories. The results of the study showed a 37 percent improvement in episodic memory performance over baseline among the subjects in the study. There was also a 35 percent improvement in both short-term and long-term retention of visual information.

The prosthetic used in this experiment was designed not to replace but to supplement the hippocampus. The study suggests that the prosthetic system can assist the hippocampus and other brain structures mediating memory when they are not functioning at optimal levels. It is a way of improving impaired memory encoding and retrieval but not restoring these capacities to predisease or preinjury levels.

Like other brain implants, the efficacy of the hippocampal prosthetic depends on a certain level of intact natural hippocampal function. Relying on this function, the system's multi-input mechanism could modulate and improve memory encoding. Its multi-output mechanism could modulate and improve memory retrieval from short- and long-term information stores. Significantly, the ability of the system to correct mistakes in neural firing patterns and how they regulate information processing enables not just the formation of new memories. It also enables the formation of memories that are accurate representations of a person's actual experience. It thus offers a way of preventing false memories in both impaired and normal memory systems. Presumably, a hippocampal prosthetic could do this without interfering with the normal process of updating the content of veridical episodic memories. By modulating neural firing patterns involved in recall of episodic memories, the prosthetic might be able to prevent hyperactivation of retrieval mechanisms and thus prevent hyperthymesia. By ensuring optimal informational input in encoding episodic memory and optimal informational output in retrieving this memory and using it for working and prospective memory, the prosthetic could benefit people with different types of declarative memory impairment.

Because it would be implanted and activated in the brain, a hippocampal neural prosthetic would involve many of the same risks as DBS. Implanting the system in the brain would be an invasive procedure, and these procedures entail significant risks. These include bleeding and infection from implantation, and unintended expanding effects of neural firing patterns on normally functioning neural circuits. These effects could destabilize this functioning if increased excitatory activity was not balanced by inhibitory activity. The design of the device and neuroimaging-guided implantation could reduce these risks, as would the ability to record brain activity in real-time during and immediately after implantation.

Like DBS, a hippocampal prosthetic would also be susceptible to brainjacking. Hackers could disrupt the mechanisms through which the system identified and modified neural firing patterns in encoding and retrieval. This interference could prevent people from forming or retrieving memories or make it difficult for them to perform these neural and mental acts. It could also have the opposite effect of hyperactivating retrieval mechanisms and generating a flood of irrelevant information in the brain. These potential outcomes of brainjacking could harm people by disrupting the mental states necessary for identity and agency. It could exacerbate amnesia or hyperthymesia by disrupting the mechanisms of a prosthetic designed to improve memory and resolve these disorders of memory capacity. Hacking a hippocampal neural prosthetic could prevent the effect it was intended to produce. The absence of external interference would not exclude the possibility of prosthetic malfunction. This may occur because the research into these systems and their effects in the human brain is at a very early stage. If the prosthetic lost its ability to process information according to the algorithm, then this could allow the generation of false memories. It could also block encoding and retrieval of veridical memories.

More human trials are needed to accurately assess the safety and efficacy of the hippocampal prosthetic. Protecting the integrity and security of the implant from external interference would be a necessary regulatory component of these trials. Consistent with the bioethical principle of nonmaleficence, this protection would be one of the ethical obligations of investigators conducting them. This obligation would obtain while research subjects have the prosthetic implanted in their brains, which could extend beyond the period of a clinical trial. It would apply to device-makers as well. Before a clinical trial testing the prosthetic in humans began, device-makers would have an obligation to build security mechanisms into these systems to reduce the risk of brainjacking, They may not be able to eliminate this risk, but they would have a duty to design the system with inbuilt obstacles to external interference. Addressing this and other risks requires multidisciplinary collaboration among biomedical engineers, neurosurgeons, neurologists and neuropsychologists specializing in memory to ensure the scientific integrity, safety and efficacy of the prosthetic. The composition of research teams often includes all of these professionals.

The study conducted by Hampson and coinvestigators involved patients with epilepsy. The DBS study on memory enhancement conducted by Suthana and coinvestigators also involved patients with this neurological disorder. Temporal lobe epilepsy may cause anterograde and retrograde amnesia in some patients. Clinical trials testing a hippocampal neural prosthetic on patients with amnesia from traumatic brain injury and neurological and neurodegenerative diseases could provide a more comprehensive view of its therapeutic potential. These trials would provide the empirical data necessary to determine the extent to which a hippocampal or other neural prosthetic could ameliorate amnesic syndromes involving impaired capacity to form, store and retrieve episodic and semantic memories.

The results of the proof-of-concept study indicated that the MIMO model can facilitate memory encoding in hippocampi that are at least partly functional. For patients with a completely dysfunctional hippocampal complex, this type of hippocampal neural prosthetic would likely not result in improved encoding or recall. There would be no natural neurobiological base on which the artificial system could operate. It is worth exploring the possibility of a more advanced MIMO model that might completely replace the hippocampus and take over its memory encoding and retrieval functions. An artificial hippocampus would have all necessary neural firing patterns programmed into it. Device-makers could create input and output algorithms that would enable the system to encode and retrieve information from stimuli the subject perceives from her body and the external world.

A closed-loop system would allow the prosthetic to respond to changes in neural firing patterns in the hippocampus and other brain structures to which it projects (Hebb, Zhang, Mahoor et al., 2014; Potter, El-Hady and Fetz, 2014; El-Hady, 2016). This is unlike an open-loop system, where output quantity has no effect on input. In open-loop systems, electrical impulses are delivered unidirectionally from the system and independently of any feedback from changes in the brain. The action of an open-loop system is not sensitive to and does not respond to the unique and dynamically fluctuating characteristics of disease states in the brain. These states would include brain damage and dysfunction in the hippocampal complex. Open-loop systems function regardless of changes in the neural environment. In contrast, the responsive capacity of a closed-loop hippocampal prosthetic could modulate neural firing patterns involved in the formation, storage and access of information about the past. This capacity would complement the mechanisms of transcription factors in the brain by ensuring optimal levels of episodic and semantic information. Closed-loop systems could strike a balance between excitatory and inhibitory neural firing patterns to regulate informational input and output.

A neural prosthetic that completely replaced a dysfunctional hippocampus and restored encoding and retrieval functions to normal or near-normal levels could benefit people with anterograde and retrograde amnesia. It could also benefit people with disorders of working, spatial and prospective memory directly or indirectly related to the inability to learn, retain and recall information. If it functioned continuously, then the prosthetic might prevent the progressive loss of hippocampal-mediated memory in neurodegenerative diseases. It could avoid the potential psychological problems with temporary memory improvement from DBS followed by decline that I mentioned earlier. It would face neurobiological challenges, however. In addition to the risk of infection, the prosthetic would have to be biocompatible with surrounding neural tissue for it to remain functional while implanted in the brain (Kennedy, Andreasen, Bartels et al., 2011). Implanted arrays may reorganize and induce changes in this tissue. These changes may be salutary, especially if they promote neuroplasticity that could regenerate neurons and neural connections in the hippocampus.

Yet even if the MIMO model could restore the neural firing patterns lost by a dysfunctional hippocampus, this does not imply that it could also induce neurogenesis and alter the underlying pathophysiology causing memory impairment or loss. DBS might have neurogenerative effects when administered at an early disease stage. But while both DBS and a hippocampal prosthetic can improve memory, their mechanisms of action and effects on neural circuits are different. Although the research is preliminary, nothing in the hippocampal prosthetic experiments in animal models or the study involving human subjects suggests that this artificial system could be neurogenerative. In that case, any restoration of memory capacity would depend entirely on the prosthetic and how long it could operate.

A prosthetic would have to respond to internal activity in multiple cell fields in circuits with which the hippocampal-entorhinal and hippocampal-fornix circuits normally interact. Its mechanism of action would have to be sensitive to circuits in limbic and cortical regions that also regulate episodic, emotional and semantic memory. The prosthetic would have to engage long-term potentiation to facilitate synaptic connectivity between these circuits without adversely affecting procedural memory mediated by a circuit in the striatum and cerebellum. The MIMO algorithm would have to regulate interaction between the artificial implant and multiple natural neural networks. There may also be subtle differences in neural firing patterns in the hippocampus and neocortex that could affect communication between them and the sensitivity of the prosthetic in responding to changes in these patterns. Although the biomimetic model is a very-large-scale integrative model, it is not known how large its scale could be in regulating different stages of episodic memory. Depending on the level of interaction between these networks, a device that integrated structurally into the brain might not completely integrate functionally into a person's complete set of memory systems.

Moreover, the prosthetic would have to respond to sensory information from the external environment through the thalamus to boost or replicate the function of grid and place cells in the brain's spatial memory system. To enable a person with severe hippocampal damage to successfully navigate the environment, the prosthetic would have to respond to sensory information from sources external to the brain. It is not obvious how it would be programmed to have the capacity for this response. Nor does the proof-of-concept study indicate how the prosthetic would affect memory reconsolidation following retrieval. It is not known whether it would facilitate reconsolidation, which is critical for long-term memory storage and, together with retrieval, is dysfunctional in retrograde amnesia. If the prosthetic enabled retrieval but not also reconsolidation, then this would limit its restorative effect.

Another question is whether the prosthetic would enable recall only of memories formed *after* it had been implanted, or whether it would also enable recall of memories formed *before* the natural hippocampus ceased to function. The restorative effect would be more limited in the first of these two scenarios and would only partly resolve amnesia. Even partial recovery of memory capacity would require functional compatibility and connectivity between an artificial hippocampus and a natural neocortex. This is critical according to Multiple Trace Theory. Retrieval of semantic memory from the neocortex does not require the hippocampus, but retrieval of episodic memory does require it. If a person had retrograde amnesia for many years, then his inability to retrieve memories would prevent them from being reconsolidated. This would gradually weaken the memory trace. A hippocampal prosthetic could avoid this, but only after it had been implanted and activated. Older memories formed and stored before the natural hippocampus became dysfunctional may have been lost. There may not be any remaining trace to be retrieved. This is another respect in which a hippocampal prosthetic could be limited in ameliorating retrograde amnesia.

Some people with a hippocampal neural prosthetic might feel unease about the fact that an artificial device implanted in their brains was controlling their capacity for declarative memory (Clausen, 2009). Yet memory encoding operates at an unconscious level in healthy brains. How the brain receives and processes sensory information through the thalamus and sends it to the hippocampus for encoding and consolidation and then to the amygdala or neocortex for storage involves processes that are outside our awareness. This is part of the adaptive purpose of memory in preventing cognitive overload in the conscious mind. We retrieve some memories by consciously trying to recall them. But at least some of the retrieval process is not conscious.

Theoretically, it would not matter whether memory processes were maintained through natural or artificial means. If a hippocampal neural prosthetic could maintain the informational input and output necessary for all relevant memory functions and served the same adaptive purpose as a natural hippocampus, then any unease about a brain implant would be unwarranted. The prosthetic would enable rather then disable episodic and other types of memory that depend on it. It could restore memory functions impaired or lost through brain injury or disease and restore some degree of functional independence in forming and executing action plans. A person could identify the device as an integral part of her brain that allows memories to be part of her conscious and unconscious mind. She could adapt to the presence of a hippocampal prosthetic in her head. People with implantable brain–computer interface systems and artificial limbs incorporate these objects into their body schema over time. Depending on how it affected her somatosensory and proprioceptive systems, a person would likely be able to incorporate a hippocampal prosthetic into her body schema just as naturally (de Vignemont, 2011; Bashford and Mehring, 2016).

A more philosophical question is how a hippocampal neural prosthetic would affect the phenomenology of recall and the meaning a person attributes to her memories. The experience of mental time travel and autobiography are conscious properties that emerge from the integrated activity of neural networks regulating memory functions. The mental act of recalling an event from the past necessarily depends on but cannot be

explained entirely in terms of its neural correlates. A hippocampal prosthetic might replace a natural hippocampus in decoding neural firing
patterns associated with episodic and semantic memory. But it could not
decode the subjective aspect of recalling personal experiences because
these experiences are not identical to firing patterns. Episodic memory
necessarily depends on brain-level mechanisms. But its subjective aspect
resists any ontological or explanatory reductionism in mechanistic terms.

Based on the brain's ability to integrate information about the past in a
memory trace, a person constructs meaning from episodic memories by
consciously organizing them into a coherent whole and understanding
how they connect with each other. This is the product not only of the
neural correlates of memory but also of how the person interacts with the
natural and social environment. What the collective set of episodic and
semantic memories means for a person results from her lived experience
in space and time. It is unclear whether a hippocampal prosthetic could
restore and maintain the subjective, first-person aspect of episodic
memory and distinguish it from the impersonal and objective facts of
semantic memory. Neural circuits are necessary for the memory a person
needs to navigate the world. But the meaning memory has for a person
depends on how she uses it to connect her present with her past, to
imagine future situations and form and execute action plans. It cannot be
captured in a mechanistic algorithm.

A neural prosthetic would not assign meaning to the information
encoded in newly formed memories. It would encode the information
in the same value-neutral way and not assign more meaning to one
memory over another. Goal-directed behavior depends on the subject's
capacity to select some memories as more valuable or meaningful to her
than others. This selection reflects the cognitive and emotional demands
of the environment as she perceives and responds to them. The hippocampal prosthetic could not make this selection because it is not part of
its function of encoding information.

Suppose that a hippocampal prosthetic could restore the capacity to
form and retrieve episodic memories. A person with anterograde or
retrograde amnesia for many years might have difficulty adjusting cognitively and emotionally to what could be a substantial change in the
content of his mental states. Unlike a person who recovered her memory
following a period of transient global amnesia, recovering memory after a
long period of amnesia may cause some distress. The psychological
adjustment probably would not be as difficult as it would be for people
with visual prosthetics implanted after a long period of blindness, for
example (Glover, 2008, pp. 20–21; Schiller and Tehovnik, 2008). While
vision and memory are both critical functions in a person's life, the

change from blindness to vision recovery would likely have a greater impact on a person's experience than the change from amnesia to memory recovery. With a properly functioning neural prosthetic, the benefit from recovering memory would outweigh any initial distress, and the person would eventually adjust to the positive change.

For those with chronic episodic and semantic amnesia and declining memory from dementia, maintaining memory would require that the prosthetic remain functional for the remaining life of the patient. One cannot assume that the MIMO functions would operate indefinitely. Depending on design limitations and how the prosthetic processes information in the brain, it may have to be replaced every few years. This could cause another difficult period of adjustment for the implant recipient. Even if it continued to operate, neural firing patterns in regions to which the prosthetic projected might have changed. This would require changing the MIMO algorithm or implanting a new system. While a closed-loop hippocampal neural prosthetic could modulate memory processes in cases where hippocampal dysfunction was localized, it probably would not be effective beyond a certain stage of generalized cortical, limbic and subcortical neurodegeneration.

At this early stage of research and development, there is insufficient empirical data to make any definitive claims about the safety and efficacy of a hippocampal neural prosthetic. Studies of the first MIMO models indicate that they can improve some memory functions but not restore them to predisease or preinjury levels. They may ameliorate retrograde or anterograde amnesia, but only to a limited extent. An advanced prosthetic that completely replaced a dysfunctional hippocampus would have considerable therapeutic potential and might restore memory functions lost from hippocampal damage. This is still very much hypothetical. There are many open questions about how an implanted artificial biomimetic device would integrate into natural neural networks and the extent to which it could restore and regulate memory encoding and retrieval. Only further research in translational neuroscience and bioengineering from animal to human models will answer these questions.

Researchers testing a neural prosthetic in humans are obligated to obtain informed consent from competent subjects and not expose them to more than a minimal risk of harm from any adverse effects of the device. This obligation is consistent with general bioethical principles of respect for persons, nonmaleficence and beneficence in medical research involving humans (US Department of Health, Education and Welfare, 1978; Emanuel, Grady, Crouch et al., 2008; Beauchamp and Childress, 2012, chs. 4–6). Informing subjects of the goals and limitations of a prosthetic would reduce the probability of unreasonable expectations about potentially positive effects. Outcomes of neural prosthetic studies that fell short of these expectations could cause psychological harm to subjects in addition to their memory impairment. This assumes that they would retain enough episodic, working and prospective memory to compare their earlier expectations and later outcomes. Despite some impairment, subjects could retain enough memory to make an informed decision to participate, or decline to participate, in a trial.

Because the effects of profound amnesia on working and prospective memory would severely impair reasoning and decision-making, amnesiacs like H.M and Clive Wearing would likely lack the cognitive capacity to consent to participate in these studies. H.M. was deemed capable enough to participate in studies of his behavior to identify the brain regions mediating different types of memory. But these studies did not involve interventions in the brain. They were significantly different from studies comparing potential benefit and risk of a neural prosthetic. Surrogates in principle could give proxy consent for severely memory-impaired subjects to participate in a clinical trial testing the device. But this would require that the known risk from earlier studies was medically acceptable and that the subject would agree to participate in this research if she were able to consent on her own (Buchanan and Brock, 1990, pp. 122ff.). If there were no adverse effects from participating in a trial, then a subject with profound amnesia might not be harmed by the lack of a beneficial outcome for her. While anyone with amnesia would have an interest in memory improvement, a person with severely impaired episodic and prospective memory may be cognitively incapable of comparing the earlier impairment and interest with the later outcome of the trial. She may not be able to experience the failure of her interest to be realized in restoring her memory. This could limit the extent to which she could be harmed.

Conclusion

Disorders of memory capacity involve impairment in the brain's capacity to encode and consolidate new memories, as in anterograde amnesia. They also involve impairment in the brain's capacity to retrieve and reconsolidate existing memories, as in retrograde amnesia. Episodic and semantic amnesia disable spatial, working and prospective memory because these types of memory rely on information about events and facts stored in the brain. At the mental level, these disorders can disrupt the experience of persisting through time and the ability to form and execute action plans. Psychopharmacology can improve function in some types of memory, though studies of different drugs have shown only modest benefits. The most active area of pharmacological research in memory disorders has been in Alzheimer's disease. Despite many clinical trials, no drugs have been able to prevent memory loss or restore memory function.

Given the limited efficacy of drugs in ameliorating disorders of memory capacity, DBS may be a therapeutic alternative because of its more direct effects on neural circuits associated with memory. Studies have shown enhanced spatial and working memory from stimulating the hippocampal-entorhinal and hippocampal-fornix circuits in patients with temporal lobe epilepsy and early-stage Alzheimer's. But deep brain stimulation is an invasive procedure with risks, and these have to be weighed against the potential benefits. A hippocampal neural prosthetic supplementing a partly dysfunctional hippocampus may also improve some types of memory, though this too is an implantable system with its own risks. In cases where the hippocampal complex was completely dysfunctional, it is possible that a neural prosthetic could replace and take over its functions in encoding and retrieving episodic and semantic memory. This is a hypothetical application of a device that has only recently been tested in a human study. Questions about its safety and efficacy can be answered only after a sufficient number of placebo-controlled trials in humans with severe memory impairment have been completed.

Assessing the comparative harm of disorders of memory content and memory capacity depends on how they affect the people who have them. This assessment may influence how we weigh the potential benefits of memory modification against the risks. Although these two types of disorders affect people in different ways, they can be equally mentally disabling. A person with anterograde or retrograde episodic amnesia may be as functionally impaired as a person with PTSD. A person with an anxiety or panic disorder may be as functionally impaired as a person with semantic amnesia. If the harmful effects of the disorders are roughly the same, then the benefit–risk assessment of interventions to treat them should also be roughly the same. However, because normal memory is not disabling, interventions to enhance it are more difficult to justify if the benefit is only moderate and they entail significant risks.

I discussed some legal aspects of dementia and anesthesia awareness in Chapters 2 and 3. Memory capacity has a broader range of legal implications. Some people may forget performing criminal actions because of amnesia. This raises the question of whether they can be responsible for them. It may depend on the extent of amnesia, or whether it is transient or chronic. Memory capacity also plays a role in criminal negligence for omissions caused by forgetting earlier actions. In cases where there is substantial loss of episodic memory, it may not be justifiable to hold a person responsible or punish her for an earlier crime if she is no longer the same person who committed it. Memory content is legally significant as well. Questions about the reliability of this content can influence assessments of testimony about criminal actions in courts of law. If memory erasure becomes feasible, then we need to consider whether a victim of a crime would have the right to erase a memory of it or an obligation to retain the memory to testify against the perpetrator. I discuss these and other legal issues regarding episodic memory in Chapter 6.

6   Legal Issues Involving Memory

The capacity to form and retrieve episodic memories is an important dimension of the criminal law. Determining that a person has been or is impaired in this capacity is critical for judgments of responsibility, mitigation or excuse for actions and omissions. Some people may not recall performing a criminal act. But memory lapses are not necessarily mitigating or excusing. These judgments depend on the intentionality of the agent at the time of action and whether memory failure is symptomatic of impaired reasoning and decision-making. Amnesia may or may not be relevant to these assessments. Episodic memory and its effects on working and prospective memory are especially pertinent to legal assessments of omissions. The inability to use information from episodic memory in forming, sustaining and executing intentions in actions may explain failures to act.

In assessing criminal responsibility for omissions, it may not be clear whether the person lacked the capacity to recall an earlier action and foresee the probable consequences or had this capacity but failed to exercise it. It can be difficult to ascertain whether a person charged with criminal negligence causing death failed to prevent the outcome because of amnesia about an earlier action or because he was not attentive to the circumstances of the action. These questions pertain to the cognitive control a person has of his behavior and the role of episodic, working and prospective memory in this control.

I discussed the role of memory in identity and the moral and legal force of precedent autonomy in Chapter 2. The connection between memory and identity is also relevant to normative assessments of criminal behavior. A person who commits a crime may subsequently undergo profound memory changes from neurodegeneration. She may not only forget committing the crime but also become a qualitatively different person from the perpetrator. Could she be held responsible and punished many years after the crime? Does criminal responsibility diminish over time? If it does, then is it because of diminished memory? The gradual weakening of episodic memory and its transition from a detailed to a generalized representation of past events can influence the reliability of eyewitness testimony in courts of law. This informs the question of whether a victim of a crime would have a right to erase a memory of his experience or an obligation to retain it to testify and serve the public interest in avoiding additional harm.

In this chapter, I examine some of the legal implications of memory capacity and incapacity. I argue that one can have some memory loss over time but remain the same person and be held responsible for one's earlier actions. Amnesia following an action does not entail that the agent had no cognitive or volitional control and was not responsible when he acted. Responsibility can transfer from an earlier to a later time. But an individual who had undergone a substantial identity change from extensive memory loss could not be held responsible and punished if he became a different person. I then consider dissociative disorders involving amnesia and whether they cause an agent to fail to satisfy the *mens rea* test in the criminal law. The core question is not whether the dissociative state caused a person to forget having performed an action. Rather, the question is whether this state impaired the agent's working and prospective memory and made him unable to form and translate an intention into a criminal act. Dissociation is a matter of degree. A person in a dissociative state may have enough working and prospective memory to act intentionally and be at least partly responsible for his behavior.

Omissions involving criminal negligence causing death may be due to failures of retrieval of an episodic memory of an earlier action. Negligence assumes that the person was able to retrieve the memory but failed to do so and in turn failed to prevent the harmful outcome. I discuss problems in distinguishing between memory lapses caused by neurological or mental disorders beyond one's control and lapses due to failed attention within one's control. In addition, I explore how a hippocampal neural prosthetic could influence these assessments. If the implant functioned normally, then presumably any claims about uncontrolled memory lapses due to neurological or psychological factors would not be defensible. Yet if the device malfunctioned and the agent could not retrieve a memory of an earlier action, then she could not be criminally negligent for the consequences of that action. I also argue that the harm a victim of a crime experiences from a memory of it and the fallibility of memory give the victim a right to erase the memory. She would not have an obligation to retain the memory to testify against the perpetrator of the crime.

I conclude by discussing the influence of neuroscience and particularly neuroimaging in assessing the role of memory in criminal law. I note the potential applications and limitations of neuroimaging in achieving a better understanding of how persons form, store and retrieve memories and how these processes influence their behavior.

Memory and Criminal Responsibility for Actions

Vernon Madison murdered police officer Julius Schulte in Mobile, Alabama, in 1985. Madison was convicted and scheduled for execution by lethal injection. In early 2018, the US Supreme Court issued a stay of execution (*Madison v. Alabama*, 2018). Madison has dementia after a series of ischemic strokes and cannot remember committing the crime. The main legal question in this case is whether the execution of someone who cannot understand the reasons for their execution violates the US Constitution's Eighth Amendment ban on cruel and unusual punishment.

There are in fact three questions pertinent to this case. The first question is whether the person's failure to remember the act indicates that he was unable to meet the cognitive and volitional conditions for criminal responsibility at the time of the act according to the M'Naghten Rule and Model Penal Code (*M'Naghten's Case*, 1843/1975, 718, 722; Model Penal Code, 1985, 401). The second question is whether a person with a neurological disorder like dementia has the capacity to understand the reasons for execution or other forms of punishment for a criminal act. The third question is whether the change in his mental states from 1985 to 2018 caused a change in identity and made him a different person from the one who committed the crime.

What I have called "cognitive" and "volitional" conditions are often described in the legal literature as "cognitive" and "control" tests. Both the cognitive capacity to recognize and respond to reasons for and against certain actions and the volitional capacity to act without coercion or compulsion are necessary for one to be liable to a judgment of criminal responsibility. Madison's subsequent dementia does not answer the first question because he was not demented when he killed Schulte. Just because a person cannot recall an action does not imply that he did not perform it intentionally and was not responsible for performing it at that time. Memory impairment can be a mitigating or excusing condition regarding an action when it interferes with the capacity to recognize and respond to reasons for or against actions.

Failure to recall a criminal act can influence a legal assessment of the act only when it is symptomatic of a memory disorder that impairs the agent's mental and physical ability to meet *mens rea* and *actus reus* requirements for state of mind and conduct in the criminal law (Model Penal Code, 1985, 2.01, 4.01). The first involves the capacity for reasoning and decision-making, while the second involves the capacity for voluntary bodily movements. Working memory necessary for decision-making relies on a store of episodic memory. But not all episodic memories are necessary for working memory. A person may fail to retrieve one or more episodic memories without this interfering with her ability to make decisions. Chronic retrograde amnesia may impair this ability. Yet a single episode of forgetting an action is not necessarily indicative of impaired working memory at the time of action. Except possibly in absence or generalized tonic-clonic seizures, a single instance of failing to access a memory is not necessarily symptomatic of a memory disorder or impaired agency. This failure may be due to a psychogenic lapse, in which case it may be difficult to determine whether the agent was unable to access the memory or was able but failed to do it. Functional neuroimaging may confirm differences in blood flow and metabolic activity between the prefrontal cortex mediating working memory, as well as other neocortical regions, and the hippocampal complex mediating episodic memory retrieval.

Brain activity can change over time, however. This is one of the limitations of using neuroimaging to establish correlations between later neural activity and earlier mental states leading to actions. Behavioral evidence may be a more reliable measure of the connection between working and episodic memory and agency. Still, this would not provide a satisfactory explanation of every action or omission.

Locke defined "person" as a "forensic" term used in "appropriating actions and their merit" (Locke, 1690/1975, Book II, ch. xxvii). He distinguished "person" from the biological term "human being." On a narrow reading of Locke, if a person cannot remember performing an action, then she is not the same person as the one who performed it. Insofar as identity is necessary for responsibility, and forgetting entails a loss of identity, the person who forgets performing an action cannot be responsible for it. But if one defines personal identity in terms of psychological connectedness and continuity, if memory is part of these psychological relations, and if they hold to varying degrees, then failing to recall one action would not disrupt identity. If identity is not disrupted, then one instance of forgetting does not absolve a person of responsibility.

Identity is relevant to holding a person responsible at a later time for an action she performed at an earlier time. It is not relevant to holding her responsible for the action when she performed it. These two issues need to be distinguished. Only if a failure to recall the action was symptomatic of a chronic disorder of episodic, working and prospective memory that precluded cognitive and volitional control when she acted would it be relevant to the question of responsibility for that action. In a singular instance of not remembering having performed an action, there would be no disruption of psychological connectedness and continuity. The agent would also be cognitively capable of foreseeing that she would be held responsible later for the action. Because of the retention of identity and cognitive control, responsibility could transfer from the earlier to the later time. An individual experiencing these lapses would remain the same person. She could justifiably be held criminally responsible if she met the *mens rea* test when she acted.

Memory loss in dementia and severe anterograde and retrograde amnesia would disrupt psychological connectedness and continuity and cause a substantial change in identity. This change could weaken the reasons for holding an offender responsible or punishing him for a crime committed in the past. Even if a person intentionally performed the act and could foresee the probable future consequences of his action, it seems unjustifiable that a person who had become demented and could not recall most of his previous actions could continue to be held criminally responsible. The dementia would have disrupted the psychological relations necessary for him to remain the same person. The transfer principle holds when memory is sufficiently integrated to maintain identity. It can accommodate some retrograde amnesia without violating the unity of one's total set of episodic memories. But this principle fails to hold when there is substantial loss of memory resulting in a substantial change in identity. Depending on the extent to which memory and other cognitive functions fail with age, one may *be* responsible for an action when one performed it but not *held* responsible for it later.

The effects of severe memory loss ground deontological reasons against holding a demented offender criminally responsible and liable to punishment. These reasons would pertain to whether he was deserving of punishment that was proportional to the crime. One could not say that the demented person deserved to be punished for his crime because his dementia made him incapable of understanding the reasons for punishment. This was the main argument in the stay of execution for Vernon Madison. Moreover, justification of punishment presupposes that the person punished is identical to the person who committed the crime. It would not be justified because dementia disrupted most of the connections between his mental states in 1985 and his mental states in 2018. The individual on death row is a different person from the person who committed the crime because of the substantial changes in his memory and other cognitive functions.

There may be a consequentialist reason for releasing Madison from continued detention because he no longer poses the same threat to the public. Dementia would cause him to lose the capacity for intentional criminal acts. Indeed, his thoughts may be so disorganized and his reasoning so diminished that they undermine his capacity for any goal-directed behavior. Nevertheless, impulsivity associated with dementia could pose a risk of harm to himself and others. This could provide a consequentialist reason for continued incarceration or, preferably, admission to a long-term care facility (Fazel, McMillan and O'Donnell, 2002). The rationale would be to prevent harm. Unlike the deontological issue about justification of punishment, the consequentialist issue about preventing harm would not rely on the question of identity. The claim that Vernon Madison in 2018 was not the same person as Vernon Madison in 1985 would not influence reasons for continued incarceration or long-term care. Deontological reasons would be relevant to the question of which situation at the end of his life would be more likely to respect his dignity and intrinsic worth as a person.

Transient global amnesia leaves a psychological gap for the person between onset and resolution. A person experiencing an episode of TGA does not recall anything from it. Yet it is not the inability to recall events that would excuse her from responsibility for any actions she committed during the episode. It is only if the amnesia prevented her from accessing the episodic, working and prospective memory necessary for voluntary agency that she would be excused.

Memory impairment is not always a mitigating or excusing condition. One could voluntarily perform an action causing impaired memory and other cognitive functions resulting in criminal behavior. A typical example is a person charged with criminal negligence causing death when driving under the influence of alcohol. The driver would be criminally responsible for his actions even if he could not recall getting into his car and driving. Responsibility would transfer from the earlier time, when he could foresee the probable consequences of driving while intoxicated, to the later time when the consequences occurred. This claim rests on diachronic agency, where the content of responsibility – what one is responsible for – extends beyond the time of action. The claim also assumes that the driver's identity does not change in the period between his earlier action and its later consequences.

There may be cases in which one could be responsible for an action one could not recall, even with a substantial change in identity. An example would be voluntarily taking a psychotropic drug that caused permanent changes in one's thought and behavior. If the agent knew the probable consequence of taking the drug, then he could be held responsible after the change for a criminal action he performed before the change. He could also be responsible for actions he performed after taking the drug. Responsibility would fall within the known risk of taking it. These attributions would be warranted despite the failure to recall what he did and the change in identity. This example is different from the dementia that made Vernon Madison unable to recall his criminal act and caused a substantial change in his identity. In my hypothetical case, the identity change is self-induced and within the agent's control. In Vernon Madison's case, the identity change is caused by a disease beyond his control.

There is the metaphysical question of making sense of holding a person responsible when "person" may refer to different individuals at different times. The extent of memory impairment is crucial to answering this question. There is also the practical question of what sorts of legal judgments could be made about an offender who had undergone this change and what sorts of deontological or consequentialist reasons one could provide to justify these judgments. The cases in which these questions arise may not be common. Still, they illustrate the complexity of relations between memory, identity and responsibility.

The relation between memory and responsibility is significantly different from the relation between memory and precedent autonomy discussed in Chapter 2. If one accepts the principle of precedent autonomy, then the moral and legal force of an advance directive holds when a person becomes demented and cannot remember what he expressed in it. The directive expresses the person's critical interest in events that occur in his entire life. A disruption in the connections between his earlier and later mental states and a substantial change in identity does not weaken the directive because it refers to the broader notion of his biological life and not just the narrower notion of his psychological life. The focus of the directive is on what should or should not be done to him in his last biological stage.

In contrast, judgments about criminal responsibility refer to actions that adversely affect others at all times. Except possibly for cases of self-induced identity change, responsibility entails that the psychological relations necessary for identity hold to a certain degree. Judging a person responsible for a criminal act presupposes that she is the same person who committed the act. This requires a certain degree of general episodic memory, even if it does not include a specific memory of an action she performed at an earlier time. In cases of dementia with amnesia about committing a criminal act, there is no justification for continuing to hold one responsible for that act. One's demented state precludes understanding the wrongfulness of the act, and one is no longer the same person.

Dissociative States and Amnesia

Amnesia is often a symptom of dissociative states. In the criminal law, it may or may not influence judgments of responsibility for actions.

Stephen Morse explains that, "in dissociative states, consciousness is not fully integrated because the normal ability self-consciously to observe oneself, to be aware of and monitor oneself, is missing or severely diminished" (Morse, 1994, p. 1642). These states include sleepwalking, fugue states and severe intoxication. Individuals in dissociative states retain some level of consciousness, but the disruption can range from moderate to severe (Spiegel and Cardena, 1991). Dissociative disorders may involve failure to access episodic and working memory when acting, to hold an intention in prospective memory, to form and consolidate an episodic memory of an action or to retrieve the memory. The first two of these failures are relevant to rationality and control of an action and whether one can be responsible for it.

Morse points out that "dissociative states are often followed by amnesia, but later amnesia does not necessarily entail that the agent lacked awareness or full intentionality during the conduct. One may be fully aware of conduct and later amnesic, and dissociated conduct may or may not be followed by amnesia" (Morse, 1994, pp. 1641–1642, note 146). Morse focuses on amnesia *following* action. The key question regarding criminal responsibility is whether amnesia *during* the performance of the action interferes with the agent's intentionality and control necessary to be responsible for it. The capacity to access information about the past and use it in working and prospective memory is critical for agency. Accessing consciousness to guide actions is closely connected with accessing episodic, working and prospective memory in forming, holding and translating intentions into actions (Block, 1995, p. 227). By itself or combined with amnesia, impaired access consciousness in dissociative states can impair control of thought and behavior.

Dissociative amnesia used to be described as "psychogenic amnesia." An affected individual fails to remember information about her life. It can disrupt the connections between the mental capacities necessary for identity and the mental capacities necessary for agency. Dissociative amnesia may be triggered by an extreme emotional reaction to psychosocial or other stress. In the criminal law, this disorder is often contrasted with organic amnesia, where failure to recall facts and events is caused by a neurological condition such as traumatic brain injury or a neurodegenerative disease like dementia (Cima, Nijman, Merckelbach et al., 2004; Gao, 2017). Despite the categorical distinction, both types of amnesia can be described accurately as dissociative because both involve disrupted information processing and self-consciousness.

I have claimed that amnesia following a criminal act is not as important as amnesia influencing the act itself in determining responsibility for it. The first type of amnesia can make an offender unfit to stand trial and an inappropriate candidate for punishment. But it would not necessarily mean that he did not act intentionally or voluntarily and was not responsible for the action. Dissociative amnesia occurring during the action could be a mitigating or excusing condition if it meant that the offender failed to satisfy the *mens rea* requirement for criminal responsibility. A dissociative state involving impaired or no access to episodic and working memory could impair or undermine the offender's capacity to recognize and respond to reasons for and against different actions. The Model Penal Code (section 2.02) states that "a person is not guilty of an offense unless she acts purposely, knowingly, recklessly, or negligently, as the law may require with respect to each material element of the offense." It is questionable whether an agent acting in a state of dissociative automatism would fail to meet the *mens rea* requirement and be excused. "Most cases of automatism do not arise from mental disorder" (Morse, 2016, p. 238), which typically is necessary for a judgment of legal insanity (Schopp, 1991, pp. 132ff.).

Dissociative amnesia involving automatism is transient and may or may not be symptomatic of a general behavioral disorder. Somnambulism, for example, is more of a neurological than a mental disorder. It is unclear to what extent dissociation in somnambulism interferes with rational and volitional capacities. If actions performed in these states were at least partly intentional, then the agent would have some cognitive and volitional control to be at least partly responsible for them. It would be inaccurate to describe them as instances of temporary insanity.

There are degrees of cognitive and volitional control of actions performed in dissociative states. "Dissociation is a degree phenomenon … it lends itself to a continuum of moral ascription ranging from full responsibility through mitigation to full excuse, depending on the resulting level of rationality impairment" (Morse, 2016, p. 239). Given the connection between working memory and rationality, one could also say that the level of rationality impairment would depend on the level of working memory impairment. It is difficult to assess actions in these states in terms of control because one's neural and mental properties and behavior can change between the time one acts and the time one's action is assessed. Morse points out that "Retrospective mental state evaluations are difficult to make, but deciding how dissociated an agent was in the past can be fearsomely difficult" (Morse, 2016, p. 238). Neither behavioral observation nor neuroimaging after the action can confirm the extent to which the dissociative state interfered with one's reasoning when one acted. Nor could these measures determine the cause of the dissociation. Assessment of an action at a later time may not be able to account for the motivational states that led to the action at the earlier time.

Dissociative states involve abrupt changes in the integration of consciousness and memory. There is no definitive test that could confirm that a person in such a state retained some or lost the capacity for reasoning when she acted and thus should be held partly responsible or excused. The best approximation to such a test would be attention to particular features of the agent's behavior. Because dissociative states are often followed by retrograde amnesia, those who observed and could describe the events leading up to and following the dissociated agent's action would be the only possible witnesses to it. This would depend on the accuracy of their episodic memory. There could be considerable variation in assessments of behavior control and responsibility in these cases.

Somnambulism, or sleepwalking, may be the most common dissociative state. Although many researchers have described somnambulism as a disorder of arousal, genetic factors and abnormal cycles of sleep and wakefulness also explain its pathophysiology (Zadra, Desautels, Petit et al., 2013). Like other dissociative states, it involves impaired information processing in the brain and altered self-consciousness. Somnambulism involves both impaired access consciousness and impaired working memory. The main issues are how impaired one's self-awareness is, the extent of memory impairment and how impaired consciousness and memory influence judgments of responsibility, mitigation or excuse. Somnambulism as such does not imply amnesia. As Antonio Zadra and coauthors point out, some sleepwalkers remember what they have done in this state (Zadra, Desautels, Petit et al., 2013, pp. 287–288). Dissociative states are amnesic when the disrupted consciousness blocks a person's access to the different types of memory necessary to form, sustain and execute intentions in voluntary actions.

These states involve a close connection between impaired consciousness and impaired memory. The disruption and loss of behavior control may come in degrees, however, as may criminal responsibility for the behavior. A person acting in a dissociative state may retain enough working and prospective memory to form and complete an action plan. The action may be partly intentional, and the person may be partly responsible for it.

A dissociative state as such does not constitute an excusing or mitigating condition in all cases. Nor does amnesia. Self-induced amnesia from alcohol intoxication would not mitigate or excuse one from responsibility for a criminal act performed in such a state. If the drinking were voluntary, then one had control of the process resulting in amnesia and could have prevented it. This would be an example of how cognitive control in the capacity to foresee a probable sequence of events can make responsibility transfer from an earlier to a later time. It can transfer even if the sequence results in a dissociative state of impaired or no awareness (Sher, 2009). Impaired episodic and working memory affecting the agent's capacity to recognize and respond to reasons for performing certain actions and refraining from performing others would not be an excusing condition in such a case.

It is instructive to consider two actual cases of sleepwalking to illustrate the connection between consciousness and memory and its effect on assessing criminal responsibility for actions performed in dissociative states. Ivy Cogdon axe-bludgeoned her nineteen-year-old daughter to death while sleepwalking (*King v. Cogdon*, 1950; Kadish and Schulhofer, 2001, pp. 193–195). Cogdon was in a dissociative state when she acted. The jury found her not guilty of homicide because of noninsane automatism. Michael Moore explains that she was unable to access reasons for not killing her daughter. These reasons would be based partly on the content of her episodic memory. Yet Moore also points out that she performed "complex routines requiring perception and readjustment in order to reach certain goals" (Moore, 1984, p. 257). Her behavior suggested that she formed and executed an intention, even if she seemed unaware of this. It suggested that she had enough working memory to form and enough prospective memory to hold and act on the intention. Like most sleepwalkers, at some level Cogdon may have been self-aware and cognizant of her relationship to her environment. Morse comments that "the movements of the unconscious agent that cause harms appear to execute more general intentions. After all, it is implausible that the harms done are random goals … To execute a general intention requires that the agent *must* be aware at some level of the intention that she is trying to execute" (Morse, 1994, pp. 1644–1645).

Forming, holding and executing the intention in the action presupposes that at least some of Cogdon's working and prospective memory was functional. It could have been enough for her to recognize and respond to a reason to cancel the intention and refrain from killing her daughter. This account of her behavior suggests that she could meet the *mens rea* test for criminal responsibility.

Some might claim that this process is beyond the mental capacity of a person in a dissociative state. Yet if she was capable of acting intentionally to some degree, then it may be plausible to say that she had enough cognitive and volitional control of her behavior to be responsible for it. Still, Cogdon did not appear to have full access to episodic memory in recognizing that the intended victim was her daughter. Nor did she have full access to working memory in processing information providing her with reasons that would have prevented her from acting. This suggests that she lacked the control necessary to be fully responsible. But the fact that the action appeared to be at least partly intentional and that she formed, held and executed a plan of action, suggests that she should not have been excused but held partly responsible.

Answers to questions about control and responsibility for actions performed in dissociative states depend not just on descriptions but also on explanations of these actions. This requires evidence of the connections between neural states and mental states at the time of action. The connections are correlative rather than causal and thus leave some uncertainty about how they are related. Morse makes the important point that "We know many of the causes of such states, but we do not well understand the neural mechanisms" (Morse, 2016, p. 238). Until neuroscientists understand these mechanisms, the question of whether a sleepwalker with partially blocked access to the types of memory necessary to recognize and respond to reasons could be criminally responsible for her actions may not have a definitive answer. One can make this claim despite the actual legal judgment in this case.

In the early morning of May 24, 1987, Kenneth Parks rose from the couch from which he was watching television and drove 23 kilometers to the house of his parents-in-law. On entering the house, he strangled his father-in-law into unconsciousness and repeatedly stabbed his mother-in-law. He then drove to the police station and told the police that he thought that he had killed some people. Parks's episodic memory of the events that occurred that early morning was fragmented. He could recall some events but not others. He was charged with first-degree murder of his mother-in-law, who died from her stabbing injuries. Parks pleaded not guilty, claiming that he was sleepwalking. His defense claimed that his actions were the product of noninsane automatism and pointed out that he had a history of somnambulism (Broughton, Billings, Cartwright et al., 1994). He was found not guilty of the crime because his fragmented memory was consistent with sleepwalking as an excusing condition. This ruling was upheld by the Supreme Court of Canada (*R v. Parks*, 1992). Presumably, Parks's impaired episodic memory in giving an account of his behavior was evidence of cognitive and volitional impairment that prevented him from controlling his actions. His amnesia suggested that he lacked the capacity for intentionality necessary for criminal responsibility. Like Cogdon, Parks's somnambulism was a disorder of both consciousness and memory that impaired his agency.

A disruption in consciousness can result in a disruption of one or more types of declarative memory. Even with disruption in declarative memory, nondeclarative types of memory like procedural memory and priming may be intact because they operate outside awareness. Because declarative and nondeclarative memory are separable, a person may

experience disordered consciousness and declarative memory in a dissociative state like somnambulism but retain procedural memory. This may
enable him to engage in complex behaviors such as driving a car. These
behaviors may involve minimal or no intentionality and may be described
as forms of automatism.

Imaging studies of sleepwalkers appear to support the distinction
between intentional and nonintentional behavior. One study using single
photon emission computed tomography (SPECT) showed deactivation
in frontal-parietal cortices while a subject was in a dissociative state.
There was also activation in the limbic anterior cingulate cortex and
subcortical thalamus. The investigators concluded that there was a "dissociation between body sleep and mind sleep" (Bassetti, 2009, p. 113).
Generalizing, one could say that a person with somnambulism "experiences motor arousal without mental arousal" (Levy, 2014, p. 74). Motor
arousal may indicate that one's procedural memory was intact. This
could allow one to engage in complex behaviors such as driving a car.
But this type of memory is not relevant to questions about cognitive
control and responsibility because it operates outside awareness.

Some of Parks's actions leading to his criminal acts could be explained
by his intact procedural memory. This would not provide a complete
explanation, however. Driving to a specific destination like the house of
his parents-in-law would involve interaction of procedural with spatial
memory. This second type of memory may be either conscious or unconscious, depending on whether actions are performed in novel or familiar
environments. Neither procedural nor spatial memory requires intentionality. Still, these types of memory alone could not explain why Parks
got into his car, drove to his parents-in-law's house and attacked them.
Complex behaviors can be nonintentional or intentional. A series of
actions culminating in a criminal act may have both nonintentional and
intentional components.

Some behaviors defy straightforward explanations. Parks's behavior
suggested that it was not entirely automatic but at least partly intentional.
It suggested that not only unconscious procedural and spatial memory
were functional when he acted, but also some degree of conscious
working and prospective memory. If some of his working and prospective
memory was functional, and if he was able to access it to form and
execute an intention to kill his mother-in-law, then, despite the Supreme
Court of Canada's decision, Parks could have been partly responsible for
his action. Responsibility requires the volitional capacity to execute an
intention in addition to the rational capacity to form it. Although somnambulism may undermine these capacities in severe cases, it does not
necessarily undermine them in all cases. Parks's sleepwalking was a

mitigating factor in assigning responsibility to him. It is questionable whether it was or should have been an excusing factor.

Parks's fragmented episodic memory after his actions was not evidence of disrupted working and prospective memory when he acted. Inability to retrieve episodic memories does not always imply impaired capacity to process information in reasoning and decision-making. If Parks had enough working and prospective memory to plan and carry out his criminal acts, then he could have been criminally responsible for them. These are declarative types of memory and as such involve consciousness. Some might claim that, a dissociative state disrupting consciousness would disrupt them as well. Yet just as consciousness comes in degrees, so too do these two types of memory. A person with retrograde amnesia may retain some degree of working memory if it relies only on short-term episodic and semantic information. It is not clear how much working and prospective memory one needs to form and execute action plans. Nevertheless, it is possible that a person in a dissociative state with some retrograde amnesia has enough cognitive and volitional control to act intentionally and be criminally responsible for his actions.

Like Cogdon, Parks's behavior appeared to display some degree of intentionality. Whether this was enough to respond to reasons for refraining from performing their criminal acts is an open question. Yet both cases suggest that dissociative amnesia may not entirely disable the memory functions necessary for intentional actions. When some degree of working and prospective memory is functional, and one can act intentionally to some degree, one may be partly responsible for actions committed in a dissociative state.

Memory and Criminal Responsibility for Omissions

On a sweltering afternoon in June 2013, in the town of Milton, Ontario, fifty-one-year-old Leslie McDonald left her two-year-old grandson, Maximus Huyskens, inside a car for several hours. The toddler died from hyperthermia, and McDonald was charged with criminal negligence causing death (Rogers, 2013). Maximus had been left in his grandmother's care while his father was at work and his mother was at an appointment. Together with purposefulness, knowledge and recklessness, negligence is one of four culpable states of mind according to the Model Penal Code. McDonald's behavior displayed an unconscious *mens rea*. By all accounts, she was able but failed to pay attention to a situation in which she was creating a substantial risk for the child. She was criminally negligent for failing to recognize and respond to this risk by not acting to prevent the tragic outcome. In November 2013, McDonald pleaded guilty to failing to provide the necessities of life.

This judgment assumed that McDonald was capable of foreseeing and preventing the probable consequence of leaving a child in an overheated car but failed to exercise this capacity. In this and other cases of criminal negligence, cognitive control provides a person with diachronic agency that extends from actions to their consequences. It can make one responsible for consequences of actions or omissions even if one is mentally incapacitated when they occur (Sifferd, 2016). McDonald's case is similar to other cases of children dying from hyperthermia because of a negligent parent or grandparent. They all appear to involve a memory lapse in failing to recall an event or action and failing to prevent a harmful outcome because of the lapse. These omissions may be caused by an inability or failure to retrieve memories. They may also be caused by an inability or failure to form and consciously hold a memory of an action plan until executing it. Although most cases of criminal negligence involve a lapse of episodic memory, they may also involve lapses of working and prospective memory.

A judgment of criminal negligence may indicate that one was able but failed to retrieve an episodic memory of an action or other event and respond accordingly. Negligence may also indicate that one was able but failed to hold an intention to act in prospective memory and execute it in the appropriate action. Encoding or forming episodic memories is beyond one's conscious control. But holding an intention to act later and retrieving a memory of an earlier action may be within one's control if one can consciously attend to the circumstances of action. Even with retrieval, in some cases it can be difficult to determine whether a person had the mental capacity to recall an action but failed to exercise it or was unable to recall it because of a neurological or mental disorder. This distinction is critical to judgments of criminal negligence.

There are different possible explanations of Leslie McDonald's behavior. One is that she believed that leaving Maximus in the car did not create a substantial risk. Given his age and the high temperature, she did not duly consider the risk of death. She would be negligent on these grounds regardless of her memory capacity. It is also possible that her lapse was symptomatic of retrograde amnesia and dysfunctional memory retrieval causing her to forget her earlier action. Another explanation is that her behavior was symptomatic of anterograde amnesia and dysfunctional memory formation. If she could not form a memory of leaving her grandson in the car, then there would be no memory of her action to retrieve. Still another explanation is that she formed a memory of her action and an intention to return to the car within a short period. But she was unable to hold the intention long enough to realize it in returning in time to prevent Maximus's death. This would involve a failure of prospective and working memory.

When memory is a factor in judgments of criminal negligence, the assumption is that the agent had the capacity to retrieve an episodic memory of an action or other event but failed to exercise it. "Failure" suggests that this mental act was to some extent within the agent's conscious cognitive control. Yet if Leslie McDonald's omission was a symptom of neural or mental dysfunction in forming, holding and retrieving memories, then it might not have been within her control. Dysfunction, and control, can hold to varying degrees. If it was beyond her control to recall her earlier action of leaving her grandson in the car or complete a plan to return to it within a brief period, then she should have been excused for her omission and his death. Diachronic agency can make responsibility transfer from an earlier time of mental capacity to a later time of mental incapacity. But it would be unreasonable to expect anyone with retrograde or anterograde amnesia to foresee the possibility of a child's death from hyperthermia and take preemptive action. If the parents had known that the grandmother had chronic retrograde or anterograde amnesia, then they would be negligent for leaving Maximus in her care and criminally responsible for the outcome.

A claim that McDonald was overloaded with information while performing multiple cognitive tasks and forgot about Maximus would not be a convincing excuse by any reasonable person standard. A claim that her memory was impaired by the heat would not be convincing either. Extreme heat and other circumstantial factors may have distracted her and impaired her reasoning, but not to the point of causing her to forget. The level of stress an agent experiences in situations involving multiple tasks may support a judgment of diminished responsibility. But the expectation of attention and care and the potential harm from the omission in this case would not support this judgment (Murray, Murray, Stewart et al., 2019). It would instead support the judgment of criminal negligence.

No structural or functional images of McDonald's brain were taken or introduced by the defense as evidence to support a claim of a neurological disorder as the cause of her memory lapse. There were no indications that the negligence was due to dementia or chronic retrograde or anterograde amnesia. Transient global amnesia might explain her lapse, but other symptoms would be necessary to confirm it. Even if her brain had been scanned with CT, MRI, PET or fMRI, there might not have been any signs of widespread gray-matter degeneration characteristic of a neurological disorder that would substantially impair reasoning and decision-making.

Advanced imaging might detect disrupted synaptic connectivity in cortical-hippocampal networks of the negligent offender's brain that ordinarily mediate memory retrieval (Tulving and Craik, 2000, pp. 448–452; Watrous, Tandon, Connor et al., 2013). This brain abnormality could be caused by a viral infection or ischemic stroke. Or it could be an early sign of dementia in someone who was asymptomatic except for an occasional memory lapse. A defensible claim that one should be excused from the charge of criminal negligence because of memory impairment would depend on a consistent chronic pattern of forgetting. Sporadic memory lapses would not provide convincing evidence of a memory disorder impairing agency. Behavioral evidence could be combined with imaging showing degeneration and reduced metabolic activity in brain regions associated with memory retrieval. Dysfunction in the hippocampal-cortical network regulating retrieval might explain problems with retrieval and memory lapses. Because Leslie McDonald's brain was not scanned, though, this account is speculative. If imaging showed that she had this abnormality, then theoretically it could provide grounds for excuse or at least question the charge of criminal negligence.

Still, imaging showing dysfunction in brain regions mediating memory consolidation and retrieval would not determine that she lacked the capacity to recall leaving her grandson in the car. Nor would it determine that she had this capacity but failed to exercise it. The fact that her reasoning seemed intact in all other respects suggests that any brain abnormalities would be subtle and probably would not add any significant information to the behavioral evidence. This would depend on the expert testimony of a neuroscientist explaining how abnormalities in the hippocampal complex and its projections to cortical areas might impair memory retrieval. Neuroimaging does not yet have the level of temporal and spatial resolution to confirm that dysfunction in these regions can explain memory lapses and that a person can be excused from negligence because of them. Advanced imaging might be able to detect brain activity that could distinguish cases in which one was unable to retrieve a memory from cases in which one was able but failed to retrieve it. This would depend on the severity of dysfunction in brain regions mediating different memory stages.

Suppose that McDonald had retrograde and anterograde amnesia from a brain injury. A hippocampal neural prosthetic was implanted in her brain to supplement her dysfunctional hippocampal complex and its connections to storage areas in the neocortex. The device could improve encoding and retrieval through its MIMO function. It could strengthen neural and mental memory functions. If the device's retrieval mechanism malfunctioned, then would this have influenced the charge of criminal negligence? It could be an excusing condition if it made her unable to recall leaving her grandson in the car. Unless this was symptomatic of a more general pattern of device malfunction, however, it would be difficult for device manufacturers and memory researchers to confirm a prosthetic retrieval failure as the cause of her memory lapse.

Again, retrospective neural and mental state evaluations are difficult to make. There may be no test confirming that there was a single device malfunction at a specific time that disrupted the process of retrieving the memory. Identifying neural firing patterns at the estimated time of malfunction could provide evidence of failure. Yet this evidence would not prove that she was unable to retrieve the memory or was able but failed to retrieve it. Because some degree of memory retrieval may be within one's conscious control, researchers could not categorically state that the lapse was caused by device malfunction.

They may be more confident in making this claim if the prosthetic did not supplement but replaced her natural hippocampal complex. If this region was completely dysfunctional, and if the prosthetic completely regulated retrieval, then failed retrieval would be due to device malfunction at that time and beyond the person's control. In that case, she could be excused for her omission.

Suppose that researchers discovered that the encoding mechanism of the hippocampal prosthetic malfunctioned. The inability to encode and consolidate new memories of actions and other events can be just as legally significant as the inability to retrieve existing memories. Indeed, encoding problems are more significant than retrieval problems because they involve the first stage of the memory process. A malfunction in the prosthetic's encoding mechanism could disable its ability to form new memories. A person could not recall an event if her brain could not form a memory of it. There would not be any content about which she could be negligent for not consciously attending to it. This is different from claims that the agent lacked the ability or failed to retrieve information about past events. A malfunction of the prosthetic's encoding mechanism would obviate the need to identify a mental lapse in retrieving the critical memory at the critical time. There would be no trace for the person to retrieve because the information constituting the trace would not have been encoded in her brain.

As with the retrieval mechanism, this judgment about the encoding mechanism would depend on whether the hippocampal prosthetic supplemented the natural hippocampus or replaced it. If the device supplemented a region that had some preserved function, then it would be difficult to identify a single instance of encoding malfunction as the cause of the omission. It would have to be symptomatic of a general pattern of device malfunction for researchers to provide a satisfactory explanation of her behavior. If the device replaced the hippocampus and completely regulated encoding, then a single encoding malfunction would be enough to explain and excuse the person for the omission. The person could not control encoding regardless of whether the device supplemented or replaced her hippocampus. But knowing the source of the encoding malfunction would be necessary to explain her behavior.

Let us assume that there is no hippocampal prosthetic implanted in the agent's brain. As explained in Chapter 2, problems with memory retrieval may result from problems with memory reconsolidation. Memories that have been consolidated and in short-term storage need to be reconsolidated following retrieval for long-term storage. Retrieval and reconsolidation are mutually influencing processes. Failed reconsolidation due to underactive LTP or protein synthesis could remove memory from long-term storage sites in the neocortex. The effect would be the same as in failed encoding because there would be no memory available for retrieval. Here too the explanation would be that the agent did not have a memory lapse because there was no memory.

Because reconsolidation operates outside awareness, failed reconsolidation could excuse the person from a charge of criminal negligence as well. However, this excuse would be more difficult to defend than an excuse based on failed encoding or retrieval, for two reasons. First, reconsolidation is not a single event but a sequence of events involving retrieving, updating and restabilizing information. It may be difficult to specify where in this sequence dysfunction occurred. Second, reconsolidation of the memory would occur after its initial encoding and consolidation and possibly after the omission. In that case, reconsolidation failure would not be an excuse for an omission more likely due to an encoding, consolidation or retrieval problem at an earlier stage of the memory process. Functional neuroimaging and behavioral observation could confirm these judgments. But a single instance of memory failure could be confirmed only within a general pattern of such failures.

Reliability of Memory in Legal Testimony

In my discussion of anesthesia awareness with recall in Chapter 3, I mentioned cases of patients hearing offensive comments about them from physicians while under general anesthesia. In some cases, physicians assault patients. These experiences can generate an emotionally charged memory and result in a psychiatric disorder such as PTSD. In Chapter 4, I pointed out that erasing memories is still experimental and investigational in humans. Let us assume that clinical trials testing drugs and neurostimulation will demonstrate the safety and efficacy of memory erasure by these methods. The question arises as to whether the victim of an assault should erase a memory of the traumatic experience. The reasons for erasure seem obvious. As a form of therapy, eliminating the memory from consciousness would eliminate the cause of a psychopathology. The victim would no longer be mentally disabled by the memory and could resume a normal mental life. Still, competing reasons for and against erasure need to be considered.

Some legal theorists have argued that there may be limits to the freedom to alter one's own brain and mind by erasing memories. A person who has been the victim of an assault and suffers from a traumatic memory of it may be obligated to retain it to testify against the perpetrator in a court of law. The obligation would be grounded in society's interest in prosecuting the offender to prevent additional public harm. As Adam Kolber states, "when it comes to balancing our freedom of memory against societal interest in public safety and prosecuting the guilty, we should not be too quick to assume that people should have unfettered rights to alter their own memories" (Kolber, 2014, p. 659; 2011; cf. Bublitz and Dresler, 2014).

Memories are one component of our mental states. The autonomy and sovereignty we have over these states imply that we have cognitive liberty to retain or remove them. They fall within the realm of our rights, specifically the negative right to noninterference in our bodies, minds and brains (Thomson, 1990, pp. 272–293). Living with a pathological memory generated by an assault is a form of interference. In principle, one should have the right to eliminate it. Yet the content of our memories may extend beyond the mind and include information about the social environment in which we act. This suggests that there may be limits to the presumed right to alter memories of traumatic events because they are too valuable to society (Kolber, 2014; Lavazza, 2015). One might agree with the claim that a witness to an atrocity or a rape victim would have an obligation to retain a memory of these events and experiences to testify against the perpetrators of these acts.

What are the limits to our freedom of memory? Where does cognitive liberty to erase memories end, and where does an obligation to retain them begin? And who decides? These questions are complicated by the fact that neuroscience has shown that episodic memories are fallible. A retrospective account of a criminal act may not be an accurate representation of what occurred. Kolber claims that even "imperfect memories can provide helpful evidence" (Kolber 2014, p. 659). But how much imperfection can the law accommodate for a memory to have probative value and be admissible in criminal court proceedings? How do we weigh the rights of individuals to control their brains and minds against the interests of society? Some would argue that a duty to protect society's interests can override a victim's right to erase a memory of a criminal act. There may be a duty to retain the memory to testify against the offender. Yet the fallibility and questionable probative value of episodic memory suggest that there might not be solid grounds for such a duty.

Forcing a crime victim to retain a traumatic memory would cause her to experience unnecessary harm in addition to the harm associated with the assault. The harm could be greater depending on the person's age. A thirty-year-old victim with a traumatic memory of her experience may suffer more than a seventy-year-old victim because the memory would persist over a longer period of her life. This would have to be weighed against the negative emotional valence of the experience, which may be stronger for the older person. Both the intensity and duration of the memory could provide reasons for erasing it. In either case, it would be unfair to the victim to expect her to experience additional actual harm from the memory to prevent potential public harm by the offender, who may or may not reoffend. Neuroimaging may predict that some offenders may commit additional crimes (Aharoni, Vincent, Harenski et al., 2013). Prediction is not deterministic but probabilistic, though, and not applicable to all types of criminal behavior.

One may appeal to John Stuart Mill's principle of liberty for guidance on this issue. Mill claims that, "over himself, over his own body and mind, the individual is sovereign" (Mill, 1859/1974, p. 69). He also claims that "the only purpose for which power can be rightfully exercised over any member of a civilized community, against his will, is to prevent harm to others" (p. 68). The first component of Mill's principle implies that one has the right to control one's thought and behavior and reject any interference in how one uses what lies within the skull. In the case of an assault victim, it is not the victim's exercise of her will in erasing a traumatic memory that poses a risk of harm to others. It is the offender who poses this risk. The harmful act was not done *by* the victim but *to* her. Because the victim is not the source but the object of harm and had her liberty violated by the offender, there would be no grounds for an obligation for her to retain the memory. Although she may exercise her right to retain the memory to testify against the offender, her equal right to erase the memory could not justifiably be overridden. The circumstances Mill envisaged as imposing limitations on individual liberty would not apply in this case.

There may be cases in which the risk and magnitude of harm to the public by the perpetrator of a criminal act would be so great that the public interest in preventing harm might override the victim's cognitive and affective liberty to erase the memory. This interest might impose an obligation on the victim to retain the memory. It would be consistent with Kolber's argument for a limited right to alter it. But these cases would be rare, and the type of case I am discussing would not be one of them. In addition, forcing the victim to retain a traumatic memory against her will for the interests of society would violate the Kantian injunction to treat people not only as means but also as ends-in-themselves (Kant, 1785/1964, p. 96). Her role as a social agent would supersede her individual right to control her brain and mind.

The fallibility of memory in eyewitness testimony further weakens claims about legal limits on altering memories. Several factors contribute to the low reliability of episodic memory in legal settings. Schacter and Loftus note that, "even under the best viewing conditions, high levels of stress can reduce an eyewitness' ability to recall and make an accurate identification" of the perpetrator of the crime (Levy, 2007, pp. 167–170; Davis and Loftus, 2012; Schacter and Loftus, 2013, p. 119). A heightened emotional response to recalling the crime can cause the release of the stress hormone cortisol. This can result in a vivid image of the act but interfere with the cognitive capacity to provide an accurate account of events surrounding it. The neurophysiology of the response can interfere with the ability to correctly identify the perpetrator. It is one example of problems with using recognition memory in criminal proceedings.

Recognition memory is divided into cognitively fast familiarity and cognitively slow recollection. Eyewitness testimony would rely on recollection as a more accurate measure of recognition. The ability to recognize faces may be due to a high level of activity in the fusiform face area in the fusiform gyrus of the inferior temporal cortex (Furl, Garrido, Dolan et al., 2011; Bilalic, 2016). Functional neuroimaging showing increased activation in this area could reliably indicate recognition and confirm accurate identification in a lineup of suspects (Gronlund, Goodsell and Andersen, 2012). Yet even if a witness's claim of facial recognition correlates with functional neuroimaging showing an active fusiform face area, he may be mistaken.

Psychological factors could confound this process despite evidence of increased fusiform activation. The negative emotional valence of remembered contextual factors can distort how one perceives and interprets faces. There may be a dissociation between subjective aspects of remembering and objective aspects of what is remembered. A person's subjective experience of recollecting an event can affect how it is recollected. This mental process may not accurately reflect objective features of the event (Kensinger, 2009; Rimmele, Davachi, Petrov et al., 2011). There may be a distortion of spatial and temporal details depending on how the amygdala interacts with the hippocampal complex and cortical regions (including the fusiform gyrus) in memory retrieval. One might assume that vivid recall of an event would ensure accuracy in describing it. But this does not necessarily follow. Elizabeth Phelps points out that "a strong emotional reaction to an event often results in an increased sense of vividness and confidence when recollecting that event in the future. However, these vivid and confident memories for emotional events can sometimes be surprisingly inaccurate in their details" (Phelps, 2012, abstract; Roediger, Wixted and DeSoto, 2012).

This does not imply that a victim's identification of the perpetrator is more often mistaken than accurate. But it shows that identification can be distorted by processes at neural and psychological levels. Or the victim may correctly identify the perpetrator but not provide an accurate account of the details and circumstances of the action. There may also be differences in the accuracy of these accounts by victims and eyewitnesses because of differences in how they experience the event. Removing the emotional content of the memory of an event might improve the accuracy of describing it. Yet this content is mainly what makes the experience wrongful and harmful to the victim. Including it in legal testimony is necessary to demonstrate the extent of the victim's harm and the criminal responsibility of the offender. But fear or other emotionally charged responses to a memory of an assault can weaken the reliability of the victim's account of it. Accuracy in recalling the criminal act and the properties that make it wrongful and harmful seem to work at cross purposes.

The reconstructive process of updating memories raises further questions about the accuracy of eyewitness testimony. This is significant because, unlike traumatic memories or retrograde amnesia, updating the content of memory is an example of how normal aspects of the memory process can affect the reliability of recalling an experience. Retaining too much information about the past may interfere with the brain's ability to promote goal-directed behavior. The details of an act performed at an earlier time, including a criminal act, may serve no purpose in our capacity to simulate future events to guide our actions. Accuracy in describing a criminal act is critical to convicting the perpetrator. But memory is not designed to provide the information necessary to make moral and legal judgments about past events. These judgments are not part of memory's adaptive purpose.

Consistent with the constructive theory of memory proposed and defended by Schacter and Addis, Joyce Lacy and Craig Stark explain that "there are adaptive reasons for generalizing and forgetting" (Lacy and Stark, 2013, p. 657). "Over time, memories typically become less episodic (highly detailed and specific) and more semantic (more broad and generalized) as the information is repeatedly retrieved and re-encoded in varying contexts" (p. 653). Through reconstruction, episodic memories of experiences may become gist memories that capture the general meaning of the event while leaving out details. Lack of sufficient detail in a memory of an assault can weaken the accuracy of an account of the event based on that memory. Accuracy can be weakened further by the fact that testimony occurs some time after the event. What might have been a clear account of the event at an earlier time may become a general and vague account of it later. Lacy and Stark state that "the belief that confident, detailed memories are always accurate and reliable is contrary to research that suggests the opposite is possible – confidently recalled recollections can sometimes be inaccurate, and real memories are not always highly confident and detailed ... especially in cases involving violence and high levels of stress" (p. 650). Memory distortion is more likely in older people. "As people age, memory for the gist of an event may remain intact, but memory for specific details of the event degrades, and individuals are more likely to falsely incorporate similar information into their memories" (Lacy and Stark, 2013, p. 652).

Lacy and Stark spell out the legal implications of fallible recollection: "The legal system needs to re-evaluate the probative value of memory. Witnessing a potentially traumatic event does not produce an unbiased, indelible memory of the event. Memory is an adaptive process based on reconstruction. It works well for what is intended – guiding current and future behavior. However, it is not infallible and therefore should not be treated as such" (p. 657). Because memory is more relevant to current and future tasks and less relevant to accounting for past events, it may limit our ability to accurately recall these events.

These considerations do not imply that eyewitness testimony based on memory should be excluded from courts of law. Phelps offers sound advice on this matter: "Learning that memories for details are often inaccurate, especially for shocking and traumatic events, may lead some to conclude that memory can never be relied on in legal proceedings. This would be a mistake, because at times memory is clearly reliable and can provide important information. Instead, legal scholars should work together with scientists to develop a better characterization of how and when memory should, or should not, be considered reliable in legal proceedings" (Phelps, 2012, p. 23). Nevertheless, memory alone cannot provide conclusive evidence of a criminal act. Other factors are necessary to confirm that the act occurred. Because of this, and because of the additional psychological harm it would cause the victim of an assault, there is no obligation for him to retain a memory of it.

There may be limits to the freedom we have to alter our memories. When there is potential for significant public harm, societal interest in public safety may override this freedom. In most cases, though, there would be no justification for limiting a person's right to erase a memory to convict the perpetrator of a criminal act. In cases where a victim's retrospective account of a criminal act had some degree of accuracy, it could secure a conviction. But she would not be obligated to retain her memory for this purpose. As an expression of her cognitive and affective liberty, she would have the right to retain the memory or erase it.

Future neuroscientific techniques might resolve the conflict between a right to erase a traumatic memory of a violent act and an obligation to retain it. Given the ability of neuroimaging and neuromodulation to target a specific engram, researchers could use them to read and save the information associated with the engram. Assuming that it was accurate, they could download the victim's memory of a traumatic event and store it in a digital repository external to the victim's brain. She could erase it from her brain after it had been stored. This process could allow the victim to unburden herself of the painful content of the memory while preserving the critical factual details. It is different from the idea of an external memory device discussed in Chapter 5 because it would not be the subject but others who would use the information for a public purpose. Although this is speculative, it is theoretically possible. Still, the process would convert episodic information about a criminal act into semantic information. It is unclear how the stored engram would retain the painful emotional content of the memory. Without this content, the wrongfulness of the act represented in the memory could be lost. Addressing the conflict in this way would fall short of reaching a just resolution to it.

Conclusion

Episodic, semantic, prospective and working memory together provide a continuous informational framework within which a person can be criminally responsible for her actions and their consequences. The cognitive control provided by memory as a reconstructive process enabling one to anticipate the future may make one responsible even when one is mentally incapacitated. What matters in these normative judgments is not whether an agent could recall performing an action, but her working and prospective memory at the time of the action. Retrograde amnesia is not necessarily indicative of impaired agency and not necessarily an excusing condition. Being in a dissociative state may be a mitigating but not always an excusing factor. Dissociative states come in degrees. In some cases, a person in such a state may retain enough cognitive and volitional control of her actions to be criminally responsible for them.

Retrograde amnesia may be an excusing condition for omissions associated with criminal negligence causing death. Yet a memory lapse would be an excusing condition only if the person lacked the cognitive capacity to retrieve a memory of an earlier action and was unable to be attentive to the circumstances surrounding his action. In these cases, memory retrieval may be at least partly within the agent's control. Anterograde amnesia may be a more likely excusing condition because the process of encoding memories occurs outside our conscious control. A person with this type of amnesia due to a neurological or mental disorder could not be responsible for failing to recall an earlier action and prevent a harmful consequence. He could not recall an action if he could not form a memory of it. Claims that retrograde amnesia can be a mitigating or excusing condition are more controversial because of uncertainty about the extent to which we can control memory retrieval.

In criminal negligence, it can be difficult to demonstrate that a person lacked the capacity to recall an earlier action, or that he had this capacity but failed to exercise it. Appeals to the person's behavior alone may not answer the question of how much control he had and whether he was negligent. Neuroimaging showing dysfunction in hippocampal-cortical pathways may clarify this question. Unless the abnormality was severe, however, imaging may not be able to demonstrate the extent of the person's memory capacity or incapacity and its effect on his actions.

Both the emotionally charged content of a memory of an assault and normal function of the memory process can weaken the reliability of memory in eyewitness testimony. This does not entirely discredit the role of memory in courts of law. But it shows that memory should be only one factor in identifying and convicting accused perpetrators of criminal acts. Advances in structural and functional brain imaging displaying differences in activation of neural networks mediating memory encoding and retrieval might distinguish true from false or imaginary memories. They may be able to distinguish more reliable from less reliable memories and contribute to more accurate assessments of assault victims' accounts. This would be a significant advance over imaging studies showing differences in brain activity mediating motor and mental arousal in somnambulism. It would directly measure activity in regions mediating episodic, working and prospective memory.

Yet it may be some time before imaging can achieve this goal. Schacter and coauthors point out that "this research is in a nascent stage ... many obstacles need to be overcome before it will be possible to seriously consider applying neuroimaging technology to courtroom cases in which the veracity of memory is at stake" (Schacter, Chamberlain, Gaesser et al., 2012, p. 249). Research subjects in imaging studies are typically college students whose brains and minds are not representative of the human population. They are often tested for the capacity to recall words, objects, shapes and faces. This recognitional capacity cannot be generalized to the more complex capacity to recall the characteristics and surrounding contextual features of actions or other events. Moreover, the blood-oxygenation-level-dependent (BOLD) signal in fMRI measuring changes in blood flow lags several seconds behind the neural activity that produces it and thus is not a direct measure of this activity (Roskies, 2013, p. 50). An additional complication is the signal-to-noise ratio in fMRI. This is a measure of how much relevant information (signal) is corrupted by junk information (noise). The ratio is too low in a single brain scan and must be averaged over many scans to have any statistical significance. Apart from these technical considerations, some subjects may be able to anticipate investigators' questions and game the system. This would undermine its scientific integrity and preclude generating reliable information about the veracity of memory (Schacter, Chamberlain, Gaesser et al., 2012, pp. 250–251).

There is also the problem of differences in brain activity at the time of a crime and the time when one testifies. Some legal theorists have pointed to these and other factors that can complicate drawing inferences from brain imaging data to the credibility of eyewitnesses' and victims' accounts of criminal acts. Even before discussion about how neuroimaging should influence assessments of testimony, there may be disagreement among neuroscientists in their interpretation of data from scans and what they might indicate about brain activity (Jones, Wagner, Faigman et al., 2013). Expressing the general view of memory researchers regarding the use of neuroimaging in the courtroom, including fMRI-based lie detection, Schacter and coauthors conclude that "while we hope that the field will make significant advances in the years to come, we believe that a cautionary stance is currently necessary in light of the present state of the art" (2012, p. 253). In the seven years since they made this comment, neuroimaging has not developed to a level suggesting that we should give up this stance.

Epilogue
The Future of Memory

Memory is generated and sustained by distributed neural networks. Declarative and nondeclarative memory are major components of our conscious and unconscious minds. While the first of these types is necessary for personal identity, both types are necessary for agency. Episodic and semantic memory are not just a recollection of past events and facts but a construction and reconstruction of them that guides our current and future behavior. Updating the content of these memories is essential for flexible thought and action in changing environments. Episodic, semantic, working and prospective memory together enable us to form and execute action plans. Procedural memory allows us to perform certain cognitive and motor tasks without having to think about performing them. This prevents the brain and mind from becoming overloaded with too much information and attending to too many tasks. The memory process thus promotes effective agency. The capacity to use information about the past to simulate possibilities and project ourselves into the future enables us to integrate our experiences into a coherent and meaningful whole. It allows each of us to construct an autobiography. This capacity is the basis of the psychological connectedness and continuity necessary for the experience of persisting through time as the same individual.

The complementary mechanisms of declarative and nondeclarative systems contribute to memory's adaptive purpose. The role of memory function and dysfunction in agency and identity has a broad range of metaphysical, ethical and legal implications. I have explored some of these implications in this book.

Disorders of memory capacity can interfere with agency and identity. Brain injury or neurodegenerative disease can cause anterograde amnesia and disrupt the encoding and consolidation of new memories. Injury and disease can also cause retrograde amnesia by disrupting retrieval and reconsolidation of existing memories. Amnesias can disrupt the capacity for reasoning and decision-making and the experience of mental time travel. Too much episodic and semantic memory may involve a different pathology and be just as maladaptive. The uncontrolled excess retrieval of information about events and facts in hyperthymesia can cause one to focus too much on the past. It can distract a person from current and future tasks and impair the ability to form and translate intentions into actions. In severe cases, exceptional recall can be a burden and even a form of suffering. Flexible and adaptive thought and behavior depend on optimal levels of encoding, consolidation, storage, retrieval and reconsolidation. Normally functioning neural networks in limbic, cortical and subcortical regions of the brain ensure a balance between learning and forgetting.

Disorders of memory content can be equally maladaptive or pathological. A persistent emotionally charged fear memory can result in the cognitive, affective and volitional impairment of psychiatric disorders including anxiety, panic, depression and posttraumatic stress disorder. Although there are differences in the underlying pathophysiology, like disorders of memory capacity disorders of memory content can interfere with reasoning and decision-making. Memory can have value or disvalue, depending on whether it enables or disables our ability to adapt to the world.

Behavioral and pharmacological therapies and neuromodulation techniques have had limited efficacy in controlling refractory memory disorders. Neuroimaging could identify the neural source of these disorders and guide interventions to treat them. Imaging has its limitations, however. In addition to problems associated with the signal-to-noise ratio, changes in blood flow measured by the BOLD signal in fMRI lag several seconds behind actual brain function causing them. Images of hemodynamic changes in fMRI and metabolic changes in PET are not literal pictures of the brain but scientific constructs from imaging data. As Tulving has noted, it is not possible to observe a specific memory trace in the brain. We infer the trace from levels of information processing associated with neural activity and synaptic connectivity. It is unclear whether imaging with higher spatial and temporal resolution will overcome the problem of inferential distance between actual brain activity and data from brain scans. Nor will advanced imaging necessarily enable researchers to distinguish between true, false and imaginary memories. Functional scans showing higher or lower levels of activation in brain regions associated with encoding and retrieval might not always be able to distinguish between memories formed and recalled through natural processes and memories generated from electrical stimulation of these regions. In the latter case, the memories may correspond to neural activity but have no connection with any external events.

Even if advanced neuroimaging could overcome these limitations, it may be some time before drugs targeting transcription factors, protein synthesis and neurotransmitters mediating memory have the degree of specificity necessary to safely and effectively improve, restore or erase it. Psychopharmacological interventions may continue to have distributed and nondiscriminating effects on other circuits. Some drugs designed to improve some memory functions might weaken others. In disorders of memory content, protein synthesis inhibitors or other drugs intended to weaken or erase pathological fear memories might also weaken or erase normal emotional memories. It is unclear whether memory-modifying psychopharmacology could be developed with the specificity necessary to have only positive and no adverse effects on memories and memory systems.

The more direct effects of high-frequency deep brain stimulation and high-intensity focused ultrasound on dysfunctional neural circuits could solve the specificity problem of psychopharmacological interventions. These techniques could improve episodic, semantic, working and spatial memory by upregulating inhibitory activity in hippocampal and neocortical circuits. They could also be used for the different purpose of weakening persistent fear memories by downregulating excitatory activity in the amygdala or other limbic circuits. The techniques could also erase these problematic memories by inducing neurophysiological silence in circuits and disrupting the information constituting the memory trace. Neuromodulation has improved spatial and working memory in some research subjects with epilepsy and others with early-stage Alzheimer's disease. It is possible that deep brain stimulation, optogenetics or other techniques could induce endogenous repair and growth mechanisms in the hippocampal-entorhinal or hippocampal-fornix circuit affected by Alzheimer's and other neurodegenerative diseases. This could slow and possibly reverse memory loss and maintain functional independence for people with these diseases. More placebo-controlled clinical trials are necessary to determine whether memory-modifying techniques could realize their therapeutic potential.

Hippocampal neural prosthetics could improve memory encoding and retrieval by supplementing a natural hippocampus damaged from injury or disease. Advanced prosthetics might not just supplement but replace damaged hippocampi and completely restore their functions. A prosthetic could restore memory encoding and retrieval through its own mechanisms and connections with other limbic, subcortical and neocortical structures. In these respects, it could ameliorate or resolve severe anterograde and retrograde amnesia. Nevertheless, a prosthetic would have to maintain the right balance between information input and output, between learning and forgetting and prevent our brains and minds from having too little or too much memory. It would have to ensure that there was an optimal level of information in the brain at any given time and optimal levels of encoding, consolidation, retrieval and reconsolidation. There are questions about how well this artificial system could integrate with natural neurons and synapses and reestablish the critical connections between neural networks to improve or restore memory and its adaptive purpose. These are hypotheses based on research that is at an early stage. Only studies involving a sufficient number of human subjects will confirm or disconfirm them.

These studies may show that the implant could seamlessly integrate into neural networks and respond precisely to neural firing patterns associated with memory. Just as people adapt psychologically to implantable brain–computer interfaces and deep brain stimulation, they could adapt to a hippocampal prosthetic. But it is questionable whether this system could capture the subjective aspects of episodic and autobiographical memory. The meaning a person attaches to these memories cannot be decoded through input and output functions or neural firing patterns of a natural or artificial hippocampus. A person constructs meaning from the conscious content of episodic memories by integrating them into a unified narrative of her unique life experience. Hippocampal and other neural networks generate and sustain memories. Yet no algorithm in a neural prosthetic and no measures of neural firing patterns could decode the meaning of memory for each person. This meaning is the product of how one perceives and responds to events and facts in the environment in which one lives and acts. A hippocampal neural prosthetic could not completely replicate a process that involves factors both inside and outside the brain. A key question is whether an artificial hippocampus, like a natural hippocampus, would be compatible or interfere with the psychological and social aspects of memory.

A more remote possibility is a neural prosthetic that could replace and restore function in declarative and nondeclarative memory systems. This would encompass not only the hippocampal complex but also the amygdala, striatum, cerebellum, frontal-temporal-parietal cortices and connections between them. Because it would involve an extensive area of the brain, there would be challenges in safely and effectively maintaining memory functions. The prosthetic would have to replicate natural memory systems that operate independently of others and those that operate interdependently. The question of the meaning of memory would arise here as well. If a very large-scale integration neural prosthetic processed information about facts and events in an objective, impersonal and value-neutral way, then episodic memories could become like semantic memories devoid of any personal significance. It is unclear how the prosthetic would affect the first-person subjective dimension of episodic memory and the role of this dimension in shaping an autobiography. The mechanism of the prosthetic could suggest how this dimension might be explained and preserved – or explained away.

Some might claim that replacing natural brain structures with hippocampal and larger-scale neural prosthetics would alter our understanding of memory, the brain–mind relation and even humanity. Whether these artificial systems blur the distinction between human and posthuman, suggest a transition from the first to the second, or suggest a different stage of humanity depends on how we conceive of ourselves. If we conceive of ourselves as essentially psychological beings, then substantial changes in our conscious and unconscious states from replacing part of a natural brain with an artificial one might indicate a transition from a human to a posthuman mode of existence. But if we conceive of ourselves as essentially biological beings constituted by integrated natural and artificial systems, then the replacement might indicate a nonsubstantial change from one phase of our existence to another. There would be no blurring of human and posthuman and no transition from one to the other but a broader concept of "human." It would not necessarily force us to adopt a different taxonomy of declarative and nondeclarative memory.

Information technology has introduced a new category of memory that is distinct from the process of constructing and reconstructing the past and the experience of mental time travel. True and false information about people on the Internet, and how it is acquired, stored and accessed, is fundamentally different from the neurobiology of encoding, consolidating and reconsolidating information. It is also fundamentally different from the psychology of retrieving and constructing meaning from it. While we have some control of our episodic and semantic memories in the private space of our brains and minds, in the public space of the cloud we have no control over the collection and dissemination of information about our lives. Technological advances may exacerbate this problem by increasing the volume of this information and the extent of its dissemination. Whereas our episodic memories gradually fade as we age, in the digital universe there is unlimited capacity to record, retain, disclose and use personal data in an impersonal way. This may allow virtual total recall of these data, though it is not just the person but also third parties who have access to it.

Psychotropic drugs and neurostimulation may erase traumatic and emotionally disturbing memories. Advanced techniques may allow researchers to download and store an engram of a traumatic experience in a digital repository. This may allow a person with this experience to erase the engram from his brain while retaining information about it for the public record. But a person cannot delete digital information about him from the external systems that store it (Mayer-Schonberger, 2011). The content of what is recorded about us does not evolve or gradually fade as our own episodic and semantic memories do. This could include downloaded and stored engrams of our experiences. One cannot clear one's browser history or indeed any digital information about oneself that lies in systems outside one's brain and mind. The line between private and public information has become blurred. We may not be able to reestablish this distinction. Digital memory will not extend our psychological or biological lives. But it could extend our biographical lives well beyond our death. The virtual immortality resulting from information about us could, in a sense, allow us to avoid oblivion. But because it could undermine the privacy and personal meaning of memory and put our biographies beyond our control, it is a form of immortality that few of us would want.

References

Abel, T., Martin, K., Bartsch, D. and Kandel, E. (1998). Memory suppressor genes: inhibitory constraints on the storage of long-term memory. *Science* 279: 338–341.

Absalom, A. and Mason, K. (eds.). (2017). *Total Intravenous Anesthesia and Target Controlled Infusions.* Berlin: Springer.

Adams, F. and Aizawa, K. (2010). Defending the bounds of cognition. In Menary, 67–80.

Addis, D. R. and Schacter, D. (2012). The hippocampus and imagining the future: where do we stand? *Frontiers in Human Neuroscience* 5: 173. doi: 103389/fnhuman.2-11.00173.

Agar, N. (2014). *Truly Human Enhancement: A Philosophical Debate of Limits.* Cambridge, MA: MIT Press.

Agren, T., Engman, J., Frick, A., Bjorkstrand, J., Larsson, E.-M., Furmark, T. et al. (2012). Disruption of reconsolidation erases a fear memory trace in the human amygdala. *Science* 337: 1550–1552.

Aharoni, E., Vincent, G., Harenski, C., Calhoun, V., Sinnott-Armstrong, W., Gazzaniga, M. et al. (2013). Neuroprediction of future re-arrest. *Proceedings of the National Academy of Sciences* 41: 6223–6228.

Akers, K., Martinez-Canabal, A., Restivo, L., Yiu, A., De Cristofaro, A., Hsiang, H.-L. et al. (2014). Hippocampal neurogenesis regulates forgetting during adulthood and infancy. *Science* 344: 598–602.

Alberini, C. (ed.). (2013). *Memory Reconsolidation.* Amsterdam: Elsevier.

Aldehri, M., Temel, Y., Alnaami, I., Jahanshahi, A. and Hescham, S. (2018). Deep brain stimulation for Alzheimer's disease: an update. *Surgical Neurology International* 9: 58. doi: 10.4103/sni.342.17.

Alkire, M., Hudetz, A. and Tononi, G. (2008). Consciousness and anesthesia. *Science* 322: 876–880.

Allison, S., Fagan, A., Morris, J. and Head, D. (2016). Spatial navigation in preclinical Alzheimer's disease. *Journal of Alzheimer's Disease* 52: 77–90.

American Psychiatric Association. (2013). *Diagnostic and Statistical Manual of Mental Disorders: DSM-5.* Washington, DC: American Psychiatric Association.

Anderson, M. (2014). Neural mechanisms of motivated forgetting. *Trends in Cognitive Sciences* 18: 279–293.

Anderson, M., Ochsner, K., Kuhl, B., Cooper, J., Robertson, E., Gabriel, S. et al. (2004). Neural systems underlying the suppression of unwanted memories. *Science* 303: 232–235.

Anonymous. (2015). On being a doctor: our family secrets. *Annals of Internal Medicine* 163: 321.

Apkarian, A., Baliki, M. and Geha, P. (2009). Towards a theory of chronic pain. *Progress in Neurobiology* 87: 81–97.

Aristotle. (1984). *On Memory and Recollection*. In *The Complete Works of Aristotle*, Revised Oxford Translation, Volume 1, trans. and ed. Jonathan Barnes. Princeton: Princeton University Press.

Augustine. (2008). *Confessions*, trans. H. Chadwick. Oxford: Oxford University Press.

Avidan, M. and Mashour, G. (2013). The incidence of intra-operative awareness in the UK: under the rate or under the radar? *Anaesthesia* 68: 334–338.

Avidan, M., Mashour, G. and Glick, D. (2009). Prevention of awareness during general anesthesia. *F1000 Medicine Reports* 1: 9. doi: 10.3410/MI-9.

Avidan, M., Jacobsohn, M., Glick, D., Burnside, B., Zhang, L., Villafranca, A. et al. (2011). Prevention of intraoperative awareness in a high-risk surgical population. *New England Journal of Medicine* 365: 591–600.

Baars, B. (1988). *A Cognitive Theory of Consciousness*. Cambridge: Cambridge University Press.

(1997). *In the Theatre of Consciousness*. New York: Oxford University Press.

Baddeley, A. (2003). Working memory: looking backward and looking forward. *Nature Reviews Neuroscience* 4: 829–839.

(2007). *Working Memory, Thought and Action*. Oxford: Oxford University Press.

Baker, L. R. (2013). *Naturalism and the First-Person Perspective*. Oxford: Oxford University Press.

Barak, S., Liu, F., Hamida, S., Yowell, Q., Neasta, J., Kharazia, V. et al. (2013). Disruption of alcohol-related memories by mTORC1 inhibition prevents relapse. *Nature Neuroscience* 16: 1111–1117.

Bartlett, F. (1932). *Remembering: A Study in Experimental and Social Psychology*. Cambridge: Cambridge University Press.

Bashford, L. and Mehring, C. (2016). Ownership and agency of an independent supernumerary hand induced by an imitation brain–computer interface. *PLoS One* 11: e0156591. doi: 10.1371/journal.pone.0156591.

Bassetti, C. (2009). Sleepwalking (somnambulism): dissociation between "body sleep" and "mind sleep." In Laureys and Tononi, 108–117.

Beauchamp, T. and Childress, J. (2012). *Principles of Biomedical Ethics*, seventh edition. New York: Oxford University Press.

Bejanin, A., Schonhaut, D., La Joie, R., Kramer, J., Baker, S., Sosa, N. et al. (2017). Tau pathology and neurodegeneration contribute to cognitive impairment in Alzheimer's disease. *Brain* 140: 3286–3300.

Belavusau, U. and Gliszczynska-Grabias, A. (eds.). (2018). *Law and Memory: Towards Legal Governance of History*. Cambridge: Cambridge University Press.

Berger, T., Hampson, R., Song, D., Goonawardena, A., Marmarelis, V. and Deadwyler, S. (2011). A cortical neural prosthesis for restoring and enhancing memory. *Journal of Neural Engineering* 8: 046017. doi: 10.1088/1741-2560/8/4/046017.

Bernecker, S. (2008). *The Metaphysics of Memory.* Dordrecht: Springer.

(2010). *Memory: A Philosophical Study.* Oxford: Oxford University Press.

Bernecker, S. and Michaelian, K. (eds.). (2017). *The Routledge Handbook of the Philosophy of Memory.* New York: Taylor & Francis.

Bernstein, D. and Loftus, E. (2009). How to tell if a particular memory is true or false. *Perspectives on Psychological Science* 4: 370–374.

Bigelow, H. (1850). Dr. Harlow's case of recovery from the passage of an iron bar through the head. *American Journal of the Medical Sciences* 20: 13–22.

Bilalic, M. (2016). Revisiting the role of the fusiform face area in expertise. *Journal of Cognitive Neuroscience* 28: 1345–1357.

Bjorkstrand, J., Agren, T., Ahs, F., Frick, A., Larsson, E.-M., Hjorth, O. et al. (2016). Disrupting reconsolidation attenuates long-term fear memory in the human amygdala and facilitates approach behavior. *Current Biology* 26: 2690–2695.

Blanke, O., Landis, T., Spinelli, L. and Seeck, M. (2004). Out-of-body experience and autoscopy of neurological origin. *Brain* 127: 243–258.

Block, N. (1995). On a confusion about a function of consciousness. *Behavioral and Brain Sciences* 18: 227–287.

(2007). Consciousness, accessibility and the mesh between psychology and neuroscience. *Behavioral and Brain Sciences* 30: 481–499.

Blumenfeld, H. (2009). The neurological examination of consciousness. In Laureys and Tononi, 15–30.

Blustein, J. (2008). *The Moral Demands of Memory.* New York: Cambridge University Press.

Borges, J. L. (1962). *Ficciones,* trans. A. Kerrigan. New York: Grove Press.

Borjigin, J., Lee, U., Liu, T., Pal, D., Huff, S., Klarr, P. et al. (2013). Surge of neurophysiological evidence and connectivity in the dying brain. *Proceedings of the National Academy of Sciences* 110: 14432–14437.

Bostrom, N. and Sandberg, A. (2009). Cognitive enhancement: methods, ethics, regulatory challenges. *Science and Engineering Ethics* 15: 311–341.

Bratman, M. (2007). *Structures of Agency: Essays.* New York: Oxford University Press.

Broca, P. (1861). Remarques sur le siege de la faculte du langage articule, suivies d'une observation d'aphemie (perte de la parole). *Bulletin de la Societe Anatomique* 6: 330–357.

Broughton, R., Billings, R., Cartwright, R., Doucette, D., Edmeads, J., Edwardh, M. et al. (1994). Homicidal somnambulism: a case report. *Sleep* 17: 253–264.

Bublitz, C. and Dresler, M. (2014). A duty to remember, a right to forget? Memory manipulations and the law. In Clausen and Levy, 1279–1307.

Bublitz, C. and Merkel, R. (2014). Crimes against minds: on mental manipulations, harms and a human right to mental self-determination. *Criminal Law and Philosophy* 8: 51–77.

Buchanan, A. and Brock, D. (1990). *Deciding for Others: The Ethics of Surrogate Decision Making.* New York: Cambridge University Press.

Bulach, R., Myles, P. and Russnak, M. (2005). Double-blind randomized controlled trial to determine extent of amnesia with midazolam given immediately before general anaesthesia. *British Journal of Anaesthesia* 94: 300–305.

Burge, T. (2003). Memory and persons. *Philosophical Review* 112: 289–337.

Buzsaki, G. and Moser, E. (2013). Memory, navigation and the theta rhythm in the hippocampal-entorhinal system. *Nature Neuroscience* 16: 130–137.

Cabeza, R., Nyberg, L. and Park, D. (eds.). (2005). *Cognitive Neuroscience of Aging: Linking Cognitive and Cerebral Aging*. New York: Oxford University Press.

Camille, N. (2004). The involvement of the orbitofrontal cortex in the experience of regret. *Science* 304: 1167–1170.

Campbell, S. (2014). *Our Faithfulness to the Past: The Ethics and Politics of Memory*, ed. C. Koggel and R. Jacobsen. New York: Oxford University Press.

Cantor, N. (2018). On avoiding deep dementia. *Hastings Center Report* 48(4): 15–24.

Cartwright, N., Cat, J., Fleck, L. and Uebel, T. (eds.). (1996). *Otto Neurath: Philosophy between Science and Politics*. Cambridge: Cambridge University Press.

Casali, A., Gosseries, O., Rosanova, M., Boly, M., Sarasso, S., Casali, K. et al. (2013). A theoretically based index of consciousness independent of sensory processing and behavior. *Science Translational Medicine* 198: 105. doi: 10.1126/scitranslmed.3006294.

Cassell, E. (2004). *The Nature of Suffering and the Goals of Medicine*, second edition. New York: Oxford University Press.

Castrioto, A., Lhommee, E., Moro, E. and Krack, P. (2013). Mood and behavioral effects of subthalamic stimulation in Parkinson's disease. *Lancet Neurology* 13: 287–305.

Cheng, J. and Ji, D. (2013). Rigid firing sequences undermine spatial memory codes in a neurodegenerative mouse model. *eLife* 2: e00647. doi: 10: 7554/wLife.00647.

Christophel, T., Klink, P., Spitzer, B., Roelfsema, P. and Haynes, J.-D. (2017). The distributed nature of working memory. *Trends in Cognitive Sciences* 21: 111–124.

Cima, M., Nijman, H., Merckelbach, H., Kremer, K. and Hollnack, S. (2004). Claims of crime-related amnesia in forensic patients. *International Journal of Law and Psychiatry* 27: 215–221.

Claparede, E. (1911). Recognition et molite. *Archives de psychologie* 11: 79–90.

Clark, A. and Chalmers, D. (1998). The extended mind. *Analysis* 58: 7–19.

Clausen, J. (2009). Man, machine and in between. *Nature* 457: 1080–1081.

Clausen, J. and Levy, N. (eds.). (2014). *Handbook of Neuroethics*. Dordrecht: Springer.

Cole-Adams, K. (2017). *Anaesthesia: The Gift of Oblivion and the Mystery of Consciousness*. Melbourne: Text Publishing.

Corkin, S. (2013). *Permanent Present Tense: The Unforgettable Life of the Amnesic Patient, H.M.* New York: Basic Books.

Craver, C., Kwan, D., Steindam, C. and Rosenbaum, R. (2014). Individuals with episodic amnesia are not stuck in time. *Neuropsychologia* 57: 191–195.

Crick, F. and Koch, C. (2003). A framework for consciousness. *Nature Neuroscience* 6: 119–126.

Damasio, A. (1992). Aphasia. *New England Journal of Medicine* 326: 531–539.

Damasio, A. and Geschwind, N. (1984). The neural basis of language. *Annual Review of Neuroscience* 7: 127–147.

Damasio, H., Grabowski, T., Frank, R., Galaburda, A. and Damasio, A. (1994). The return of Phineas Gage: clues about the brain from the skull of a famous patient. *Science* 264: 1102–1105.

Davidson, A. (2014). Fiddling with memory. *Journal of Medical Ethics* 40: 659–660.

Davidson, D. (2001). *Essays on Actions and Events*, second edition. Oxford: Clarendon Press.

Davis, D. (Deborah) and Loftus, E. (2012). Inconsistencies between law and the limits of human cognition: the case of eyewitness identification. In Nadel and Sinnott-Armstrong, 29–58.

Davis, D. (Dena). (2014). Alzheimer disease and pre-emptive suicide. *Journal of Medical Ethics* 40: 543–549.

Davis, J. (2018). *New Methuselahs: The Ethics of Life Extension*. Cambridge, MA: MIT Press.

Davis, N. and van Koningsbruggen, M. (2013). "Non-invasive" brain stimulation is not non-invasive. *Frontiers in Systems Neuroscience* 7: 76. doi: 10.3389/fnsys.2013.00076.

De Brigard, F. (2013). Is memory for remembering? recollection as a form of episodic hypothetical thinking. *Synthese* 190: 1–31.

(2017). Memory and imagination. In Bernecker and Michaelian, 127–140.

De Brigard, F., Addis, D. R., Ford, J., Schacter, D. and Giovanello, K. (2013). Remembering what could have happened: neural correlates of episodic counterfactual thinking. *Neuropsychologia* 51: 2401–2414.

Deeprose, C. and Andrade, J. (2006). Is priming during anesthesia unconscious? *Consciousness and Cognition* 15: 1–23.

DeGrazia, D. (2000). Prozac, enhancement and self-creation. *Hastings Center Report* 30(2): 34–40.

(2005). Enhancement technologies and human identity. *Journal of Medicine and Philosophy* 30: 261–283.

de Grey, A. (2004a). Escape velocity: why the prospect of extreme life extension matters now. *PLoS Biology* 2: 723–726.

(2004b). Three self-evident life-extension truths. *Rejuvenation Research* 7: 165–167.

Dehaene, S. and Naccache, L. (2001). Towards a cognitive neuroscience of consciousness: basic evidence and a workspace framework. *Cognition* 79: 1–37.

Dehaene, S. and Changeux, J.-P. (2011). Experimental and theoretical approaches to conscious processing. *Neuron* 70: 200–227.

Deisseroth, K., Etkin, A. and Malenka, R. (2015). Optogenetics and the circuit dynamics of psychiatric disease. *Journal of the American Medical Association* 313: 2019–2020.

De Lavilleon, G., Lacroix, M., Rondi-Reig, L. and Benchenane, K. (2015). Explicit memory creation during sleep demonstrates a causal role of place cells in navigation. *Nature Neuroscience* 18: 493–495.

De Marco, M. and Venneri, A. (2018). Volume and connectivity of the ventral tegmental area are linked to neurocognitive signatures of Alzheimer's disease in humans. *Journal of Alzheimer's Disease* 63: 167–180.

Demertzi, A. and Laureys, S. (2012). Where in the brain is pain? Evaluating painful experiences in non-communicative patients. In Richmond, Rees and Edwards, 89–98.

Deng, Z.-D., Lisanby, S. and Peterchev, A (2013). Electric field depth-focality trade off in transcranial magnetic stimulation: simulation comparison of 50 coil designs. *Brain Stimulation* 6: 1–13.

De Ridder, D., Vanneste, S. Gillett, G., Manning, P., Glue, P. and Langguth, B. (2016). Psychosurgery reduces uncertainty and increases free will? a review. *Neuromodulation* 19: 239–248.

De Vignemont, F. (2011). Embodiment, ownership and disownership. *Consciousness and Cognition* 20: 82–93.

Dresser, R. (2014). Pre-emptive suicide, precedent autonomy and preclinical Alzheimer disease. *Journal of Medical Ethics* 40: 550–551.

(2018). Advance directives and discrimination against people with dementia. *Hastings Center Report* 48(4): 26–27.

Dunsmoor, J., Murty, V., Davachi, L. and Phelps, E. (2015). Emotional learning selectively and retroactively strengthens memories for related events. *Nature* 520: 345–348.

Dworkin, G. (1988). *The Theory and Practice of Autonomy*. New York: Cambridge University Press.

Dworkin, R. (1993). *Life's Dominion: An Argument about Abortion, Euthanasia and Individual Freedom*. New York: Vintage Books.

Earp, B., Sandberg, A., Kahane, G. and Savulescu, J. (2014). When is diminishment a form of enhancement? rethinking the enhancement debate in biomedical ethics. *Frontiers in Systems Neuroscience* 8: 12. doi: 10.3389/fnsys.2014.00012.

Ebbinghaus, H. (1885/1913). *Memory: A Contribution to Experimental Psychology*, trans. H. Ruger and C. Bussenius. New York: Teachers College, Columbia University.

Edmonds, D. (ed.). (2016). *Philosophers Take on the World*. Oxford: Oxford University Press.

Eichenbaum, H. (2012). *The Cognitive Neuroscience of Memory: An Introduction*, second edition. New York: Oxford University Press.

(2017). The integration of space, time and memory. *Neuron* 95: 1007–1018.

Einstein, G. and McDaniel, M. (2005). Prospective memory: multiple retrieval processes. *Current Directions in Psychological Science* 14: 286–290.

El-Hady, A. (ed.). (2016). *Closed Loop Neuroscience*. Amsterdam: Elsevier.

Elias, W., Lipsman, N., Onda, W., Ghanouni, P., Kim, Y., Lee, W. et al. (2016). A randomized trial of focused ultrasound thalamotomy for essential tremor. *New England Journal of Medicine* 375: 730–739.

Elliott, C. (2000). Pursued by happiness and beaten senseless: Prozac and the American dream. *Hastings Center Report* 30(2): 7–12.

(2003). *Better than Well: American Medicine Meets the American Dream*. New York: W. W. Norton.

Elsey, J. and Kindt, M. (2016). Manipulating human memory through reconsolidation: ethical implications of a new therapeutic approach. *American Journal of Bioethics-Neuroscience* 7: 225–236.

Emanuel, E., Grady, C., Crouch, R., Lie, R., Miller, F. and Wendler, D. (2008). *The Oxford Textbook of Clinical Research Ethics.* New York: Oxford University Press.

Eriksson, J., Vogel, E., Lansner, A., Bergstrom, F. and Nyberg, L. (2015). Neurocognitive architecture of working memory. *Neuron* 88: 33–46.

Erler, A. (2011). Does memory modification threaten our authenticity? *Neuroethics* 4: 235–249.

Errando, C. and Aldecoa, C. (2014). Awareness with explicit recall during general anaesthesia: current status and issues. *British Journal of Anaesthesia* 112: 1–4.

Ezzyat, Y., Wanda, P., Levy, D, Kadel, A., Aka, A., Pedisich, I. et al. (2018). Closed-loop stimulation of temporal cortex rescues functional networks and improves memory. *Nature Communications* 9: 365. doi: 10.1038/s41467-017-027550.

Fazel, S., McMillan, J. and O'Donnell, I. (2002). Dementia in prison: ethical and legal implications. *Journal of Medical Ethics* 28: 156–159.

Feinberg, J. (1986). *Harm to Others.* New York: Oxford University Press.

Feinstein, J., Buza, C., Hurlemann, R., Follmer, R., Dahdaleh, N., Coryell, W. et al. (2013). Fear and panic in humans with bilateral amygdala damage. *Nature Neuroscience* 16: 270–272.

Fell, J., Staresina, B., DoLam, A., Widman, G., Helmstaedter, C., Elger, C. et al. (2013). Memory modulation by weak synchronous deep brain stimulation: a pilot study. *Brain Stimulation* 6: 270–273.

Ferzan, K. and Morse, S. (eds.). (2016). *Legal, Moral and Metaphysical Truths: The Philosophy of Michael Moore.* New York: Oxford University Press.

Fradera, A. and Kopelman, M. (2009). Memory disorders. In Squire, L. (ed.), *Encyclopedia of Neuroscience.* Amsterdam: Elsevier, 751–760.

Franke, A., Gransmark, P., Agricola, A., Schule, K. Rommel, T., Sebastian, A. et al. (2017). Methylphenidate, modafinil and caffeine for cognitive enhancement in chess: a double-blind randomised controlled trial. *European Neuropsychopharmacology* 27: 248–260.

Frankfurt, H. (1988a). Freedom of the will and the concept of a person. In Frankfurt, 1988c, 11–25.

  (1988b). Identification and externality. In Frankfurt, 1988c, 58–68.

  (1988c). *The Importance of What We Care About.* New York: Oxford University Press.

Frankland, P. and Josselyn, S. (2013). Neuroscience: memory and the single molecule. *Nature* 493: 312–313.

Fu, H., Rodriguez, G., Herman, M., Emrani, S., Nahmani, E., Barrett, G. et al. (2017). Tau pathology induces excitatory neuron loss, grid cell dysfunction, and spatial memory deficits reminiscent of early Alzheimer's disease. *Neuron* 93: 533–541.

Furl, N., Garrido, L., Dolan, R., Driver, J. and Duchaine, B. (2011). Fusiform gyrus face-selectivity reflects facial recognition ability. *Journal of Cognitive Neuroscience* 23: 1785–1790.

Gao, G. (2017). Amnesia and criminal responsibility. *Journal of Law and Biosciences* 4: 194–204.

Garcia Marquez, G. (1970). *One Hundred Years of Solitude*, trans. G. Rabassa. New York: Harper & Row.

Gathercole, S. (2007). Working memory: what it is, and what it is not. In Roediger, Dudai and Fitzpatrick, 155–158.

Glannon, W. (2002a). Identity, prudential concern and extended lives. *Bioethics* 16: 266–283.

(2002b). Reply to Harris. *Bioethics* 16: 292–297.

(2006). Psychopharmacology and memory. *Journal of Medical Ethics* 32: 164–168.

(2011). The neuroethics of memory. In Nalbantian, Matthews and McClelland, 233–251.

(2014a). Anesthesia, amnesia and harm. *Journal of Medical Ethics* 40: 651–657.

(2014b). Intraoperative awareness: consciousness, memory and law. *Journal of Medical Ethics* 40: 663–664.

(2018). Brain implants to erase memories. *Frontiers in Neuroscience* 11: 584. doi: 10.3389/fnins.2017.00584.

Glover, J. (2008). *Choosing Children: Genes, Disability and Design*. Oxford: Oxford University Press.

Goddard, N. and Smith, D. (2013). Unintended awareness and monitoring of depth of anaesthesia. *Continuing Education in Anaesthesia, Critical Care and Pain* 13: 213–217.

Gold, P. (2008). Protein synthesis inhibition and memory: formation versus amnesia. *Neurobiology of Learning and Memory* 89: 201–211.

Goldstein, L. and Kapur, N. (2012). Psychiatric aspects of memory disorders in epilepsy. In Zeman, Kapur and Jones-Gotman, 259–282.

Goodman, W. (2011). Electroconvulsive therapy in the spotlight. *New England Journal of Medicine* 364: 1785–1787.

Gordijn, B. and Chadwick, R. (eds.). (2009). *Medical Enhancement and Posthumanity*. London: Springer.

Graf, P. and Schacter, D. (1985). Implicit and explicit memory for new associations in normal and amnesic subjects. *Journal of Experimental Psychology: Learning, Memory and Cognition* 11: 501–518.

Graff-Radford, J. and Josephs, K. (2012). Primary progressive aphasia and transient global amnesia. *Archives of Neurology* 69: 401–404.

Gronlund, S., Goodsell, C. and Andersen, S. (2012). Lineup procedures in eyewitness identification. In Nadel and Sinnott-Armstrong, 59–83.

Haig, S. (2007). The ethics of erasing a bad memory. *Time*, October 15. http://content.time.com/time/health/article/0.8599.1671492.00html.

Hamani, C., Holtzheimer, P., Lozano, A. and Mayberg, H. (eds.). (2016). *Neuromodulation in Psychiatry*. Oxford: Wiley Blackwell.

Hampson, R., Song, D. Opris, I., Santos, L., Shin, D., Gerhardt, G. et al. (2013). Facilitation of memory encoding in primate hippocampus by a neuroprosthesis that promotes task-specific neuronal firing. *Journal of Neural Engineering* 10: 066013. doi: 10.1088/1741-2560/10/6/066013.

Hampson, R., Song, D., Robinson, B., Fetterhoff, D., Dakos, A., Roder, B. et al. (2018). Developing a hippocampal neural prosthetic to facilitate human memory encoding and recall. *Journal of Neural Engineering* 15: 036014. doi: 10.1088/1741-2552/aaaed7.

Harlow, J. (1848). Passage of an iron rod through the head. *Boston Medical Surgical Journal* 39: 389–393.

(1868). Recovery from the passage of an iron bar through the head. *Publications of the Massachusetts Medical Society* 2: 327–347.

Harris, J. (2002). A response to Walter Glannon. *Bioethics* 16: 284–291.

(2007). *Enhancing Evolution: The Ethical Case for Making Better People.* Princeton: Princeton University Press.

Hassabis, D. and Maguire, E. (2007). Deconstructing episodic memory with construction. *Trends in Cognitive Sciences* 11: 299–306.

Hassabis, D., Kumaran, D., Vann, S. and Maguire, E. (2007). Patients with hippocampal amnesia cannot imagine new experiences. *Proceedings of the National Academy of Sciences* 104: 1726–1731.

Hasselmo, M. (2009). A model of episodic memory: mental time travel along encoded trajectories using grid cells. *Neurobiology of Learning and Memory* 92: 559–573.

(2012). *How We Remember: Brain Mechanisms of Episodic Memory.* Cambridge, MA: MIT Press.

Hebb, A., Zhang, J., Mahoor, M., Tsiokos, C., Matlack, C., Chizeck, H. et al. (2014). Closing the feedback loop: closed-loop neurostimulation. *Neurosurgery Clinics of North America* 25: 187–204.

Henry, M., Fishman, J. and Youngner, S. (2007). Propranolol and the prevention of post-traumatic stress disorder: is it wrong to erase the "sting" of bad memories? *American Journal of Bioethics* 7(9): 12–20.

Hudetz, A. and Pearce, R. (eds.). (2009). *Suppressing the Mind: Anesthetic Modulation of Memory and Consciousness.* New York: Humana Press.

Hyman, S. (2005). Addiction: a disease of learning and memory. *American Journal of Psychiatry* 162: 1414–1422.

Illes, J. and Sahakian, B. (eds.). (2011). *Oxford Handbook of Neuroethics.* Oxford: Oxford University Press.

Inman, C., Manns, J., Bijanki, K., Bass, D., Hamann, S., Drane, D. et al. (2018). Direct electrical stimulation of the amygdala enhances declarative memory in humans. *Proceedings of the National Academy of Sciences* 115: 98–103.

Insel, T. (2010). Rethinking schizophrenia. *Nature* 468: 187–193.

Irish, M. and Piguet, O. (2013). The pivotal role of semantic memory in remembering the past and imagining the future. *Frontiers in Behavioral Neuroscience* 7: 27. doi: 10.3389/fnbeh.2013.00027.

Irish, M., Addis, D. R., Hodges, J. and Piguet, O. (2012). Considering the role of semantic memory in episodic future thinking: evidence from semantic dementia. *Brain* 135: 2178–2191.

Iuculano, T. and Cohen Kadosh, R. (2013). The mental cost of cognitive enhancement. *Journal of Neuroscience* 33: 4482–4486.

Izquierdo, I., Cammarota, M., Medina, J. and Bevilaqua, L. (2004). Pharmacological findings on the biochemical bases of memory processes: a general view. *Neural Plasticity* 11: 159–189.

James, W. (1890). *Principles of Psychology,* Volume 1. New York: Henry Holt.

Jaworska, A. (1999). Respecting the margins of agency: Alzheimer's patients and the capacity to value. *Philosophy & Public Affairs* 28: 105–138.

Jones, A. (2013). Toronto anesthesiologist guilty on all counts of molesting women during surgery. *The Globe and Mail*, November 19. www.theglobeandmail.com/news/national/toronto-anesthesiologist-found-guilty-of-sexual-assault/article15 504798/.

Jones, O., Wagner, A., Faigman, D. and Raichle, M. (2013). Neuroscientists in court. *Nature Reviews Neuroscience* 14: 730–736.

Jongsma, K., Kars, M. and van Delden, J. (2019). Dementia and advance directives: some empirical and normative concerns. *Journal of Medical Ethics* 45: 92–94.

Josselyn, S. (2010). Continuing the search for the engram: examining the mechanism of fear memories. *Journal of Psychiatry and Neuroscience* 35: 221–228.

Jotterand, F. and Dubljevic, V. (eds.). (2016). *Cognitive Enhancement: Ethical and Policy Implications in International Perspectives*. New York: Oxford University Press.

Juengst, E. (1998). What does enhancement mean? *In Parens*, 29–47.

Jun, Y., Duffy, J. and Josephs, K. (2013). Primary progressive aphasia and apraxia of speech. *Seminars in Neurology* 33: 342–347.

Kadish, S. and Schulhofer, S. (2001). *Criminal Law and Its Processes: Cases and Materials*, eighth edition. New York: Aspen.

Kamm, F. M. (2007). *Intricate Ethics: Rights, Responsibilities and Permissible Harm*. Oxford: Oxford University Press.

Kandel, E. (2001). The molecular biology of memory storage: a dialogue between genes and synapses. *Science* 294: 1030–1038.

Kensinger, E. (2009). Remembering the details: effects of emotion. *Emotion Review* 1: 99–113.

Kant, I. (1785/1964). *Groundwork of the Metaphysics of Morals*, trans. H. J. Paton. New York: Harper & Row.

Kass, L. (2003). *Beyond Therapy: Biotechnology and the Pursuit of Happiness*. New York: Harper Collins.

Kennedy, P., Andreasen, D., Bartels, T., Ehirim, P., Mao, H., Velliste, M. et al. (2011). Making the lifetime connection between brain and machine for restoring and enhancing function. *Progress in Brain Research* 194: 1–25.

Kent, C., Mashour, G., Metzger, N., Posner, K. and Domino, K. (2013). Psychological impact of unexpected and explicit recall of events occurring during surgery under sedation, regional anaesthetics and general anaesthesia: data from the Anaesthesia Awareness Registry. *British Journal of Anaesthesia* 110: 381–387.

Kerssens, C. and Alkire, M. (2010). Memory formation during general anesthesia. In Mashour, 47–73.

Keven, N., Kurczek, J., Rosenbaum, R. and Craver, C. (2018). Narrative construction is intact in episodic amnesia. *Neuropsychologia* 110: 104–112.

Kihlstrom, J. and Cork, R. (2007). Consciousness and anesthesia. In Velmans and Schneider, 628–639.

Kim, W. and Cho, J. (2017). Encoding of discriminative fear memory by input-specific LTP in the amygdala. *Neuron* 95: 988–990.

Kindt, M., Soeter, M. and Vervliet, B. (2009). Beyond extinction: erasing human fear responses and preventing the return of fear. *Nature Neuroscience* 12: 256–258.

King v. Cogdon (1950). Supreme Court of Victoria, Australia.

Kitamura, T., MacDonald, C. and Tonegawa, S. (2015). Entorhinal-hippocampal neuronal circuits bridge temporally discontinuous events. *Learning and Memory* 22: 438–443.

Kitamura, T., Ogawa, S., Roy, D., Okuyama, T., Morrissey, M., Smith, L. et al. (2017). Engrams and circuits crucial for systems consolidation of a memory. *Science* 356: 73–78.

Klein, S. and Nichols, S. (2012). Memory and the sense of personal identity. *Mind* 121: 677–702.

Klingberg, T. (2008). *The Overflowing Brain: Information Overload and the Limits of Working Memory*, trans. N. Betteridge. New York: Oxford University Press.

Knight, R. and Eichenbaum, H. (2013). Multiplexed memories: a view from human cortex. *Nature Neuroscience* 16: 257–258.

Koch, C. (2012). *Consciousness: Confessions of a Romantic Reductionist*. Cambridge, MA: MIT Press.

Koch, C. and Mormann, F. (2010). The neurobiology of consciousness. In Mashour, 24–46.

Kolber, A. (2006). Therapeutic forgetting: the legal and ethical implications of memory dampening. *Vanderbilt Law Review* 59: 1561–1626.

(2008). Freedom of memory today. *Neuroethics* 1: 145–148.

(2011). Neuroethics: give memory-altering drugs a chance. *Nature* 476: 275–276.

(2014). The limited right to alter memory. *Journal of Medical Ethics* 40: 658–659.

Kopelman, M. (2002). Disorders of memory. *Brain* 125: 2152–2190.

Kroes, M., Tendolkar, I., Van Wingen, G., Van Waarde, J., Strange, B. and Fernandez, G. (2014). An electroconvulsive therapy procedure impairs reconsolidation of episodic memories in humans. *Nature Neuroscience* 17: 204–206.

Krugers, H., Zhou, M., Joels, M. and Kindt, M. (2011). Regulation of excitatory synapses and fearful memories by stress hormones. *Frontiers in Behavioral Neuroscience* 5: 62. doi: 10.3389/fnbeh.2011.00062.

Lacagnina, A., Brockway, E. Crovetti, C., Shue, E., McCarty, M., Sattler, K. et al. (2019). Distinct hippocampal engrams control extinction and relapse of fear memory. *Nature Neuroscience* 22: doi: 10.1038/s41593-019-0361-z

Lacy, J. and Stark, C. (2013). The neuroscience of memory: implications for the courtroom. *Nature Neuroscience* 14: 649–658.

Langsjo, J., Alkire, M., Kaskinoro, K., Hayama, H., Maksimow, A., Kaisti, K. et al. (2012). Returning from oblivion: imaging the neural core of consciousness. *Journal of Neuroscience* 32: 4935–4943.

LaPointe, L. and Stierwalt, J. (2018). *Aphasia and Related Neurogenic Language Disorders*. fifth edition. New York: Thieme.

Lattal, K. and Wood, M. (2013). Epigenetics and persistent memory: implications for reconsolidation and silent extinction beyond the zero. *Nature Neuroscience* 16: 124–129.

Laureys, S. and Tononi, G. (eds.). (2009). *The Neurology of Consciousness: Cognitive Neuroscience and Neuropathology*. Amsterdam: Elsevier.

Lavazza, A. (2015). Erasing traumatic memories: when context and social interests can outweigh personal autonomy. *Philosophy, Ethics and Humanities in Medicine* 10: 3. doi: 10.1186/s13010-014-0021-6.

(2016). What we may forget when discussing human memory manipulation. *American Journal of Bioethics-Neuroscience* 7: 249–251.

Laxton, A., Tang-Wai, D., McAndrews, M., Zumsteg, D., Wemberg, R., Keren, R. et al. (2010). A phase 1 trial of deep brain stimulation of memory circuits in Alzheimer's disease. *Annals of Neurology* 68: 521–534.

LeDoux, J. (2007). Consolidation: challenging the traditional view. In Roediger, Dudai and Fitzpatrick, 171–176.

(2015). *Anxious: Using the Brain to Understand and Treat Fear and Anxiety*. New York: Viking.

Lee, A., Kanter, B., Wang, D., Lim, J., Zou, M. Qiu, C. et al. (2013). PRKCZ null mice show normal learning and memory. *Nature* 493: 416–419.

Lee, J., Nader, K. and Schiller, D. (2017). An update on memory reconsolidation updating. *Trends in Cognitive Sciences* 21: 531–545.

Leichsenring, F. and Leweke, F. (2017). Social anxiety disorder. *New England Journal of Medicine* 376: 2255–2264.

LePort, A., Stark, S., McGaugh, J. and Stark, C. (2015). Highly superior autobiographical memory: quality and quantity of retention over time. *Frontiers in Psychology* 6: 17. doi: 10.3389/psyq.2015.02017.

Levy, N. (2007). *Neuroethics: Challenges for the 21st Century*. Cambridge: Cambridge University Press.

(2014). *Consciousness and Moral Responsibility*. Oxford: Oxford University Press.

Lewis, D. (1976). Survival and identity. In Rorty, 17–40.

Li, F. and Tsien, J. (2009). Memory and the NMDA receptors. *New England Journal of Medicine* 361: 302–303.

Liao, S. M. and Sandberg, A. (2008). The normativity of memory modification. *Neuroethics* 1: 85–99.

Liao, S. M., Sandberg, A. and Savulescu, J. (2016). Should we be erasing memories? In Edmonds, 232–235.

Lipsman, N., Schwartz, M., Huang, Y., Lee, I., Sankar, T., Chapman, M. et al. (2013). MR-guided focused ultrasound thalamotomy for essential tremor; a proof-of-concept study. *Lancet Neurology* 12: 462–468.

Lisman, J. and Fallon, J. (1999). What maintains memories? *Science* 283: 339–340.

Locke, J., (1690/1975). *An Essay Concerning Human Understanding*, ed. P. H. Nidditch. Oxford: Clarendon Press.

Lonergan, M., Oliveira-Figueroa, L., Pitman, R. and Brunet, A. (2013). Propranolol's effects on the consolidation and reconsolidation of long-term emotional memory in healthy participants: a meta-analysis. *Journal of Psychiatry and Neuroscience* 38: 222–231.

Lozano, A. and Lipsman, N. (2013). Probing and regulating dysfunctional circuits using deep brain stimulation. *Neuron* 77: 406–424.

Lozano, A., Fosdick, L., Chakravarty, M., Leoutsakos, J. M., Munro, C., Oh, E. et al. (2016). A phase II study of fornix deep brain stimulation in mild Alzheimer's disease. *Journal of Alzheimer's Disease* 54: 777–787.

Luria, A. R. (1969). *The Mind of a Mnemonist*. London: Jonathan Cape.

Lynch, G. and Gall, C. (2006). Ampakines and the three-fold path to cognitive enhancement. *Trends in Neurosciences* 29: 554–562.

Madison v. Alabama (2018). 17–7505 (Capital Case).

Mander, B., Rao, V., Lu, B., Saletin, J., Lindquist, J., Ancoli-Israel, S. et al. (2013). Prefrontal atrophy, disrupted NREM slow waves and impaired hippocampal-dependent memory in aging. *Nature Neuroscience* 16: 357–364.

Mandler, G. (1980). Recognizing: the judgment of previous occurrence. *Psychological Review* 87: 252–271.

Manning, L., Cassel, D. and Cassel, J.-C. (2013). St. Augustine's reflections on memory and time and the current concept of subjective time in mental time travel. *Behavioral Sciences* 3: 232–243.

Mansour, A., Farmer, M., Baliki, M. and Apkarian, A. (2014). Chronic pain: the role of learning and brain plasticity. *Restorative Neurology and Neuroscience* 32: 129–139.

Maquet, P. (2000). Sleep on it! *Nature Neuroscience* 3: 1235–1236.

Mashour, G. (ed.). (2010). *Consciousness, Awareness and Anesthesia*. New York: Cambridge University Press.

Mashour, G. and Alkire, M. (2013). Consciousness, anesthesia and the thalamocortical system. *Anesthesiology* 118: 13–15.

Mashour, G. and Avidan, M. (2015). Intraoperative awareness: controversies and non-controversies. *British Journal of Anaesthesia* 112: i20–i26.

Massimini, M. and Tononi, G. (2018). *Sizing Up Consciousness: Towards an Objective Measure of the Capacity for Experience*. trans. F. Anderson. Oxford: Oxford University Press.

Matthews, P. (2011). The mnemonic brain: neuroimaging, neuropharmacology, and disorders of memory. In Nalbantian, Matthews and McClelland, 99–127.

Mayer-Schonberger, V. (2011). *Delete: The Virtue of Forgetting in the Digital Age*. Princeton: Princeton University Press.

Maylor, E., Chater, N. and Brown, G. (2001). Scale invariance in the retrieval of retrospective and prospective memories. *Psychonomic Bulletin & Review* 8: 162–167.

McDaniel, M., Umanath, S., Einstein, G. and Waldrum, E. (2015). Dual pathways to prospective remembering. *Frontiers in Human Neuroscience* 9: 392. doi: 10.3389/fnhum.2015.00392.

McGaugh, J. (2000). Memory: a century of consolidation. *Science* 287: 248–251.
(2004). The amygdala modulates the consolidation of memories of emotionally arousing experiences. *Annual Review of Neuroscience* 27: 1–28.
(2015). Consolidating memories. *Annual Review of Psychology* 66: 1–24.

McKerlie, D. (2013). *Justice between the Young and the Old*. New York: Oxford University Press.

Medford, N., Phillips, M., Brierly, B., Brammer, M., Bullmore, E. and David, A. (2005). Emotional memory: separating content and context. *Psychiatry Research: Neuroimaging* 138: 247–258.

Mele, A. (1995). *Autonomous Agents: From Self-Control to Autonomy*. New York: Oxford University Press.

Meloni, E., Gillis, T., Manoukian, J. and Kaufman, M. (2014). Xenon impairs reconsolidation of fear memories in a rat model of posttraumatic stress disorder (PTSD). *PLoS ONE* 9: e106189. doi: 10.1371/journal. pone.0106189.

Menary, R. (ed.). (2010). *The Extended Mind*. Cambridge, MA: MIT Press.

Merkow, M., Burke, J., Ramayya, A., Sharan, A., Sperling, M. and Kahana, M. (2017). Stimulation of the human medial temporal lobe between learning and recall selectively enhances forgetting. *Brain Stimulation* 10: 645–650.

Metzinger, T. and Hildt, E. (2011). Cognitive enhancement. In Illes and Sahakian, 245–264.

Mhuircheartaigh, R., Warnaby, K., Rogers, R., Jbabdi, S. and Tracey, I. (2013). Slow-wave activity saturation and thalamocortical isolation during propofol anesthesia in humans. *Science Translational Medicine* 5: 208ra148. doi: 10.1126/scitranslmed.3006007.

Michaelian, K. (2016). *Mental Time Travel: Episodic Memory and Our Knowledge of the Personal Past*. Cambridge, MA: MIT Press.

Mill, J. S. (1859/1974). *On Liberty*, ed. G. Himmelfarb. London: Penguin.

Miller, D., Dresser, R. and Kim, S. (2019). Advance euthanasia directives: a controversial case and its ethical implications. *Journal of Medical Ethics* 45: 84–89.

Milner, B. (1959). The memory defect in bilateral hippocampal lesions. *Psychiatric Research Reports of the American Psychiatric Association* 11: 43–58.

Milner, B. and Penfield, W. (1955–1956). The effect of hippocampal lesions on recent memory. *Transactions of the American Neurological Association* 20: 42–48.

M'Naghten's Case. (1843/1975). 8 Eng. Rep. London: Her Majesty's Stationery Office.

Model Penal Code. (1985). Official Draft and Commentaries. Philadelphia: American Law Institute.

Monfils, M. and Holmes, E. (2018). Memory boundaries: opening a window inspired by reconsolidation to treat anxiety, trauma-related, and addiction disorders. *Lancet Psychiatry* 4: 1032–1042.

Moore, M. (1984). *Law and Psychiatry: Rethinking the Relationship*. New York: Cambridge University Press.

Morse, S. (1994). Culpability and control. *University of Pennsylvania Law Review* 142: 1587–1660.

(2016). Moore on the mind. *In Ferzan and Morse*, 233–249.

Morse, S. and Roskies, A. (eds.). (2013). *A Primer on Criminal Law and Neuroscience*. New York: Oxford University Press.

Moscovitch, M. (2007). Memory: why the engram is elusive. In Roediger, Dudai and Fitzpatrick, 17–21.

(2012). Memory before and after H.M.: an impressionistic historical perspective. In Zeman, Kapur and Jones-Gotman, 19–50.

Moscovitch, M. and Nadel, L. (1998). Consolidation and the hippocampal complex revisited: in defense of the multiple trace model. *Current Opinion in Neurobiology* 8: 297–300.

Moser, E. and Moser, M.-B. (2008). A metric for space. *Hippocampus* 18: 1142–1156.

Moser, M.-B., Rowland, D. and Moser, E. (2015). Place cells, grid cells and memory. *Cold Spring Harbor Perspectives in Biology* 7: e021808. doi: 10.1101/cshperspect.a021808.

Murray, S., Murray, E., Stewart, G., Sinnott-Armstrong, W. and De Brigard, F. (2019). Responsibility for forgetting. *Philosophical Studies* 176: 1177–1201. doi: 10.1007/s11098-018-1053-3.

Nadel, L. (2007). Consolidation: the demise of the fixed trace. In Roediger, Dudai and Fitzpatrick, 177–182.

Nadel, L. and Moscovitch, M. (1997). Memory consolidation, retrograde amnesia and the hippocampal complex. *Current Opinion in Neurobiology* 7: 217–227.

Nadel, L. and Sinnott-Armstrong, W. (eds.). (2012). *Memory and Law*. New York: Oxford University Press.

Nadel, L., Hupbach, A., Gomez, R. and Newman-Smith, K. (2012). Memory formation, consolidation and transformation. *Neuroscience and Biobehavioral Reviews* 36: 1640–1645.

Nader, K. (2013). The discovery of memory reconsolidation. In Alberini, 1–13.

Nader, K. and Einarsson, E. (2010). Memory reconsolidation: an update. *Annals of the New York Academy of Sciences* 1191: 27–41.

Nader, K., Schafe, G. and LeDoux, J. (2000). Fear memories require protein synthesis in the amygdala for reconsolidation after retrieval. *Nature* 406: 722–726.

Nader, K., Hardt, O., Einarsson, E. and Finnie, P. (2013). The dynamic nature of memory. In Alberini, 15–41.

Nalbantian, S., Matthews, P. and McClelland, J. (eds.). (2011). *The Memory Process: Neuroscientific and Humanistic Perspectives*. Cambridge, MA: MIT Press.

Naqvi, N., Rudrauf, D., Damasio, H. and Bechara, A. (2007). Damage to the insula disrupts addiction to cigarette smoking. *Science* 315: 531–534.

Neurath, O. (1921). *Anti-Spengler*. Munich: Callwey Verlag.

Nicolas, S. (1996). Experiments on implicit memory in a Korsakoff patient by Claparede (1907). *Cognitive Neuropsychology* 13: 1193–1199.

Nielson, D., Smith, T., Sreekumar, V., Dennis, S. and Sederberg, P. (2015). Human hippocampus represents space and time during retrieval of real-world memories. *Proceedings of the National Academy of Sciences* 112: 11078–11083.

Nietzsche, F. (1872–1874/1995). On the utility and liability of history for life. In *Unfashionable Observations*, Volume 2, trans. R. Gray. Stanford: Stanford University Press.

Nitsche, M., Boggio, P., Fregni, F. and Pascual-Leone, A. (2009). Treatment of depression with transcranial direct current stimulation (tDCS): a review. *Experimental Neurology* 219: 14–19.

Northoff, G. (2014). *Minding the Brain: A Guide to Philosophy and Neuroscience*. London: Palgrave Macmillan.

Ocampo, A., Reddy, P., Martinez-Redondo, P., Platero-Luengo, A., Hatanaka, F., Hishida, T. et al. (2016). In vivo amelioration of age-associated hallmarks by partial reprograming. *Cell* 167: 1719–1733.

O'Keefe, J. and Nadel, L. (1978). *The Hippocampus as a Cognitive Map*. Oxford: Oxford University Press.

O'Keefe, J. and Burgess, N. (1996). Geometric determinants in the place fields of hippocampal neurons. *Nature* 381: 425–428.

Olshansky, S. J. and Carnes, B. (2001). *The Quest for Immortality: Science at the Frontiers of Aging.* New York: W. W. Norton.

Olshansky, S. J., Hayflick, L. and Carnes, B. (2002). Position statement on human aging. *The Journal of Gerontology* 57: B292–B297.

Olson, E. (1997). *The Human Animal: Personal Identity without Psychology.* New York: Oxford University Press.

(2007). *What Are We? A Study in Personal Ontology.* New York: Oxford University Press.

Owen, A. and Coleman, M. (2008). Functional imaging in the vegetative state. *Nature Reviews Neuroscience* 9: 235–243.

Owen, A., Coleman, M., Boly, M., Davis, M., Laureys, S. and Pickard, J. (2006). Detecting awareness in the vegetative state. *Science* 313: 1402.

Palombo, D., Alain, C., Soderlund, H., Khuu, W. and Levine, B. (2015). Severely deficient autobiographical memory (SDAM) in healthy adults: a new mnemonic syndrome. *Neuropsychologia* 72: 105–118.

Pandharipande, P., Girard, T., Jackson, J., Morandi, A., Thompson, J., Pun, B. et al. (2013). Long-term cognitive impairment after critical illness. *New England Journal of Medicine* 369: 1306–1316.

Pandit, J., Cook, T., Jonker, W. and O'Sullivan, E. (2013). A national survey of anaesthetists (NAP5 Baseline) to estimate an annual incidence of accidental awareness during general anaesthesia in the UK. *British Journal of Anaesthesia.* 110: 501–509.

Parens, E. (ed.). (1998). *Enhancing Human Traits: Ethical and Social Implications.* Washington, DC: Georgetown University Press.

Parfit, D. (1984). *Reasons and Persons.* Oxford: Clarendon Press.

(2012). We are not human beings. *Philosophy* 87: 5–28.

Park, D. and Gutchess, A. (2005). Long-term memory and aging: a cognitive neuroscience perspective. In Cabeza, Nyberg and Park, 218–245.

Parker, E., Cahill, L. and McGaugh, J. (2006). Case of unusual autobiographical memory. *Neurocase* 12: 35–49.

Parnia, S., Spearpoint, K., de Vos, G., Fenwick, P., Goldberg, D., Yong, J. et al. (2014). AWARE – AWAreness during REsuscitation: a prospective study. *Resuscitation* 85: 1799–1805.

Parsons, R. and Ressler, K. (2013). Implications of memory modulation for post-traumatic stress and fear disorders. *Nature Neuroscience* 16: 146–153.

Patihis, L., Frenda, S., LePort, A., Petersen, N., Nichols, R., Stark, C. et al. (2013). False memories in highly superior autobiographical memory individuals. *Proceedings of the National Academy of Sciences* 110: 20947–20952.

Patterson, D. and Pardo, M. (eds.). (2016). *Philosophical Foundations of Law and Neuroscience.* Oxford: Oxford University Press.

Penfield, W. (1952). Memory mechanisms. *Archives of Neurology and Psychiatry* 67: 178–198.

(1975). *Mystery of the Mind: A Critical Study of Consciousness and the Human Brain.* Princeton: Princeton University Press.

Penfield, W. and Boldrey, E. (1937). Somatic motor and sensory representation in the cerebral cortex of man as studied by electrical stimulation. *Brain* 60: 389–443.

Penfield, W. and Milner, B. (1958). Memory deficit produced by bilateral lesions in the hippocampal zone. *AMA Archives of Neurology and Psychiatry* 79: 475–497.

Penfield, W. and Rasmussen, T. (1950). *The Cerebral Cortex of Man: A Critical Study of Localization of Function.* New York: Macmillan.

Phelps, E. (2004). Human emotion and memory: interactions of the amygdala and hippocampal complex. *Current Opinion in Neurobiology* 14: 198–202.

(2012). Emotion's impact on memory. In Nadel and Sinnott-Armstrong, 7–26. doi:10.1093/acprof:050/9780199920754.023.000.

Pitman, R. (2011). Will reconsolidation blockade offer a novel treatment for posttraumatic stress disorder? *Frontiers in Behavioral Neuroscience* 5: 11. doi: 10.3389/fnbeh.2011.00011.

(2015). Harnessing reconsolidation to treat mental disorders. *Biological Psychiatry* 78: 819–820.

Pitman, R., Sanders, K., Zusman, R., Healy, A., Cheema, F., Lasko, N. et al. (2002). Pilot study of secondary prevention of posttraumatic stress disorder with propranolol. *Biological Psychiatry* 51: 189–192.

Plato. (1962). *Collected Dialogues*, ed. E. Hamilton and H. Cairns. Princeton: Princeton University Press.

Pliny. (1942). *Natural History*, Volume 2, trans. H. Rackham. Cambridge, MA: Harvard University Press.

Ploghaus, A., Tracey, I., Gati, J., Clare, S., Menon, R., Matthews, P. et al. (1999). Dissociating pain from its anticipation in the human brain. *Science* 284: 1979–1981.

Posner, J., Saper, C., Schiff, N. and Plum, F. (2007). *Plum and Posner's Diagnosis of Stupor and Coma*, fourth edition. New York: Oxford University Press.

Potter, S., El-Hady, A. and Fetz, E. (2014). Closed loop neuroscience and neuroengineering. *Frontiers in Neural Circuits* 8: 115. doi: 10.3389/fncir.2014.00115.

President's Council on Bioethics (US). (2003). Staff working paper: Better memories? The promise and perils of pharmacological interventions. March 6, session 4. www.bioethics.giv/transcripts/mar03.html/.

Price, D. (2000). Psychological and neural mechanisms of the affective dimension of pain. *Science* 288: 1769–1772.

Price, J. (with Davis, B.). (2008). *The Woman Who Can't Forget.* New York: Free Press.

Pryor, K., Root, J., Mehta, M., Stern, E., Pan, H., Veselis, R. et al. (2015). Effect of propofol on the medial temporal lobe emotional memory system: a functional magnetic resonance imaging study in human subjects. *British Journal of Anaesthesia* 115: i104–i113.

Pycroft, L., Boccard, S., Owen, S., Stein, J., Fitzgerald, J., Green, A. et al. (2016). Brainjacking: implant security issues in invasive neuromodulation. *World Neurosurgery* 92: 454–462.

Quiroga, R. (2012). *Borges and Memory: Encounters with the Human Brain*, trans. J.P. Fernandez. Cambridge, MA: MIT Press.

R v. Parks (1992). 2 S.C.R. 871. Supreme Court of Canada.

Rascovsky, K., Growden, M., Pardo, I., Grossman, S. and Miller, B. (2009). "The quicksand of forgetfulness": semantic dementia in *One Hundred Years of Solitude. Brain* 132: 2609–2616.

Rees, G. and Edwards, S (2009). Is pain in the brain? *Nature Clinical Practice Neurology* 5: 76–77.

Reinhart, R. and Nguyen, J. (2019). Working memory revived in older adults by synchronizing rhythmic brain circuits. *Nature Neuroscience* 22. doi: 10.1038/s41593-019-0371-x.

Richardson, M., Strange, B. and Dolan, R. (2004). Encoding of emotional memories depends on amygdala and hippocampus and their interactions. *Nature Neuroscience* 7: 278–285.

Richmond, S., Rees, G. and Edwards, S. (eds.). (2012). *I Know What You're Thinking: Brain Imaging and Mental Privacy*. Oxford: Oxford University Press.

Rimmele, U., Davachi, L., Petrov, R., Dougal, S. and Phelps, E. (2011). Emotion enhances the subjective feeling of remembering, despite lower accuracy for contextual details. *Emotion* 11: 4246. doi: 10.1037/a0024246.

Roache, R. (2018). What sort of person could have a radically extended lifespan? *Journal of Medical Ethics* 44: 217–218.

Roediger, H., Dudai, Y. and Fitzpatrick, S. (eds.). (2007). *Science of Memory: Concepts*. Oxford: Oxford University Press.

Roediger, H., Wixted, J. and DeSoto, K. (2012). The curious complexity between confidence and accuracy in reports from memory. In Nadel and Sinnott-Armstrong, 84–118.

Rogers, K. (2013). Grandmother charged in death of Milton, Ont. toddler left alone in hot car. *Globe and Mail*, July 5. www.theglobeandmail.com/news/national/grandmother-charged-in-death-of-milton-toddler-left-alone-in-hot-car/article/013019696.

Rolls, E. (2007). *Memory, Attention, and Decision-Making: A Unifying Computational Neuroscience Approach*. Oxford: Oxford University Press.

Rorty, A. (ed.). (1976). *The Identities of Persons*. Berkeley: University of California Press.

Rose, C. (2012). Charlie Rose, *The Brain Series: Motor Disorders*, June 15. https://charlierose.com/collecions/3/clip/15489.

Roskies, A. (2002). Neuroethics for the new millennium. *Neuron* 35: 21–23.
    (2013). Brain imaging techniques. In Morse and Roskies, 37–74.

Rowlands, M. (2010). *The New Science of the Mind: From Extended Mind to Embodied Phenomenology*. Cambridge, MA: MIT Press.
    (2017). *Memory and the Self: Phenomenology, Science and Autobiography*. New York: Oxford University Press.

Roy, D., Kitamura, T., Okuyama, T., Ogama, S., Sun, C., Obata, Y. et al. (2017). Distinct neural circuits for the formation and retrieval of episodic memories. *Cell* 170: 1000–1012.

Rugg, M. and Vilberg, K. (2013). Brain networks underlying episodic memory retrieval. *Current Opinion in Neurobiology* 23: 255–260.

Ryan, T., Roy, D., Pignatelli, M., Arons, A. and Tonegawa, S. (2015). Engram cells retain memory under retrograde amnesia. *Science* 348: 1007–1013.

Sacks, O. (2007). The abyss: music and amnesia. *The New Yorker*, September 24. www.newyorker.com/reporting/2007/09/24/070924fa_fact_Sacks.

Sacktor, T. (2011). How does PKMzeta maintain long-term memory? *Nature Reviews Neuroscience* 12: 9–15.

Sacktor, T. and Hell, J. (2017). The genetics of PKMZeta and memory mainten-ance. *Science Signalling* 10: 2327. doi: 10.1126/scisignal.ss02327.

Samuel, N., Taub, A., Paz, R. and Raz, A. (2018). Implicit aversive memory under anaesthesia in animal models: a narrative review. *British Journal of Anaesthesia* 121: 219–232.

Sandkuhler, J. and Lee, J. (2013). How to erase memory traces of pain and fear. *Trends in Neurosciences* 36: 343–352.

Sartre, J.-P. (2007). *Existentialism Is a Humanism*. New Haven: Yale University Press,

Sauchelli, A. (2018). Life-extending enhancements and the narrative approach to personal identity. *Journal of Medical Ethics* 44: 219–225.

Savica, R., Grossardt, B., Bower, J., Boeve, B., Ahlskog, J. and Rocca, W. (2013). Incidence of dementia with Lewy bodies and Parkinson's disease dementia. *JAMA Neurology* 70: 1396–1402.

Savulescu, J., Sandberg, A. and Kahane, G. (2011). Well-being and enhance-ment. In Savulescu, ter Meulen and Kahane, 3–18.

Savulescu, J., ter Meulen, R. and Kahane, G. (eds.). (2011). *Enhancing Human Capacities*. Oxford: Wiley-Blackwell.

Schacter, D. (1987). Implicit memory: history and current status. *Journal of Experimental Psychology: Learning, Memory and Cognition* 13: 501–518.

(1996). *Searching for Memory: The Brain, the Mind and the Past*. New York: Basic Books.

(2001). *The Seven Sins of Memory: How the Mind Forgets and Remembers*. Boston: Houghton Mifflin.

(2007). Memory: delineating the core. In Roediger, Dudai and Fitzpatrick, 23–27.

(2012). Adaptive constructive processes and the future of memory. *American Psychologist* 67: 603.

Schacter, D. and Addis, D. R. (2007a). The cognitive neuroscience of construct-ive memory: remembering the past and imagining the future. *Philosophical Transactions of the Royal Society B: Biological Sciences* 362: 773–786.

(2007b). Constructive memory: the ghosts of past and future. *Nature* 445: 27.

Schacter, D. and Loftus, E. (2013). Memory and the law: what can cognitive neuroscience contribute? *Nature Neuroscience* 16: 119–123.

Schacter, D. and Scarry, E. (eds.). (2000). *Memory, Brain and Belief*. Cambridge, MA: Harvard University Press.

Schacter, D., Addis, D. R. and Buckner, R. (2008). Episodic simulation of future events. *Annals of the New York Academy of Sciences* 1124: 39–60.

Schacter, D., Guerin, S. and St. Jacques, P. (2011). Memory distortion: an adaptive perspective. *Trends in Cognitive Sciences* 15: 467–474.

Schacter, D., Addis, D. R., Hassabis, D., Martin, V., Spreng, R. and Szpunar, K. (2012). The future of memory: remembering, imagining and the brain. *Neuron* 76: 677–694.

Schacter, D., Chamberlain, J., Gaesser, B. and Gerlach, K. (2012). Neuroima-ging of true, false and imaginary memories: findings and implications. In Nadel and Sinnott-Armstrong, 233–262.

Schechtman, M. (2014). *Staying Alive: Personal Identity, Practical Concerns and the Unity of a Life*. Oxford: Oxford University Press.

Schiller, D. and Phelps, E. (2011). Does reconsolidation occur in humans? *Frontiers in Behavioral Neuroscience* 5: 24. doi: 10.3389/fnbeh.2011.00024.

Schiller, D., Monfils, M.-H., Raio, C., Johnson, D., LeDoux, J. and Phelps, E. (2010). Preventing the return of fear in humans using reconsolidation update mechanisms. *Nature* 403: 49–53.

Schiller, P. and Tehovnik, E. (2008). Visual prosthesis. *Perception* 37: 1529–1559.

Schneider, G. (2010). Monitoring anesthetic depth. In Mashour, 114–130.

School, D. (2014). Pat Martino discusses relearning to play guitar after a near-fatal brain aneurysm left him with amnesia. *Lehigh Valley Live*, May 15. www.lehighvalleylive.com/music.index.ssf/2014/05/15/pat_martino_discusses relearni.html.

Schopp, R. (1991). *Automatism, Insanity and the Psychology of Criminal Responsibility*. New York: Cambridge University Press.

Schuepbach, W., Rau, J., Knudsen, K., Volkmann, J., Krack, P., Timmerman, L. et al. (2013). Neurostimulation for Parkinson's disease with early motor complications. *New England Journal of Medicine* 368: 610–622.

Scoville, W. and Milner, B. (1957). Loss of recent memory after bilateral hippocampal lesions. *Journal of Neurology, Neurosurgery and Psychiatry* 20: 11–21.

Shalev, A., Liberzon, I. and Marmar, C. (2017). Post-traumatic stress disorder. *New England Journal of Medicine* 376: 2459–2469.

Shepherd, J. (2014). Minimizing harm via psychological intervention: response to Glannon. *Journal of Medical Ethics* 40: 662–663.

Sher, G. (2009). *Who Knew? Responsibility without Awareness*. New York: Oxford University Press.

Shushruth, S. (2013). Exploring the neural basis of consciousness through anesthesia. *Journal of Neuroscience* 33: 1757–1758.

Sifferd, K. (2016). Unconscious *mens rea*: lapses, negligence and criminal responsibility. In Patterson and Pardo, 161–178.

Sillivan, S., Vaissiere, T. and Miller, C. (2015). Neuroepigenetic regulation of pathogenic memories. *Neuroepigenetics* 1: 28–33.

Soeter, M. and Kindt, M. (2015). An abrupt transformation of phobic behavior after a post-retrieval amnesic agent. *Biological Psychiatry* 78: 880–886.

Sorabji, R. (2004). *Aristotle on Memory*, second edition. Chicago: University of Chicago Press.

Spiegel, D. and Cardena, E. (1991). Disintegrated experience: the dissociative disorders revisited. *Journal of Abnormal Psychology* 100: 366–378.

Squire, L. (2004). Memory systems of the brain: a brief history and current perspective. *Neurobiology of Learning and Memory* 82: 171–177.

(2009). The legacy of patient H.M. for Neuroscience. *Neuron* 61: 6–9.

Squire, L. and Bayley, P. (2007). The neuroscience of remote memory. *Current Opinion in Neurobiology* 17: 185–196.

Squire, L. and Kandel, E. (2009). *Memory: From Mind to Molecules*, second edition. Greenwood Village: Roberts & Company.

Squire, L. and Zola, S. (1996). Structure and function of declarative and non-declarative memory systems. *Proceedings of the National Academy of Sciences* 93: 13515–13522.

Squire, L., van der Horst, A., McDuff, S., Frascino, J., Hopkins, R. and Maudlin, K. (2010). Role of the hippocampus in remembering the past and imagining the future. *Proceedings of the National Academy of Sciences* 107: 19044–19048.

Stickgold, R. (2011). Memory in sleep and dreams: the construction of meaning. In Nalbantian, Matthews and McClelland, 73–95.

Stone, C. and Bietti, L. (eds.). (2016). *Contextualizing Human Memory: An Interdisciplinary Approach to Understanding How Individuals and Groups Remember the Past.* New York: Routledge.

Strick, P., Dum, R. and Fiez, J. (2009). Cerebellum and nonmotor function. *Annual Review of Neuroscience* 32: 413–434.

Strohminger, N. and Nichols, S. (2014). The essential moral self. *Cognition* 131: 159–171.

  (2015). Neurodegeneration and identity. *Psychological Science* 26. doi: 10.1177/0956797615592381.

Suddendorf, T. and Corballis, M. (2007). The evolution of foresight: what is mental time travel, and is it unique to humans? *Behavioral and Brain Sciences* 30: 299–313.

Sulmasy, D. (2018). An open letter to Norman Cantor regarding dementia and physician-assisted suicide. *Hastings Center Report* 48(4): 28–30.

Suthana, N., Haneef, Z., Stern, J., Mukamel, R., Behnke, B., Knowlton, B. et al. (2012). Memory enhancement and deep brain stimulation of the entorhinal area. *New England Journal of Medicine* 366: 501–510.

Sutton, J. (1998). *Philosophy and Memory Traces: Descartes to Connectionism.* Cambridge: Cambridge University Press.

Tang, Y.-P., Shimizu, E., Dube, G., Rampon, C., Kerchner, G., Zhou, M. et al. (1999). Genetic enhancement of learning and memory in mice. *Nature* 401: 63–69.

Tasbihgou, S., Vogels, M. and Absalom, A. (2018). Accidental awareness during general anaesthesia: a narrative review. *Anaesthesia* 73: 112–122.

Taylor, C. (1992). *The Ethics of Authenticity.* Cambridge, MA: Harvard University Press.

Tenenbaum, E. and Reese, B. (2007). Memory-altering drugs: shifting the paradigm of informed consent. *American Journal of Bioethics* 7(9): 40–42.

Terrace, H. and Metcalfe, J. (eds.). (2005). *The Missing Link in Cognition: Origins of Self-Reflective Consciousness.* Oxford: Oxford University Press.

Thompson, R. and Madigan, S. (2005). *Memory: The Key to Consciousness.* Princeton: Princeton University Press.

Thomson, J. J. (1990). *The Realm of Rights.* Cambridge, MA: Harvard University Press.

Tomlinson, S., Davis, N., Morgan, H. and Bracewell, R. (2014a). Cerebellar contributions to spatial memory. *Neuroscience Letters* 578: 182–186.

  (2014b). Cerebellar contributions to verbal working memory. *Cerebellum* 13: 354–361.

Tononi, G. and Koch, C. (2008). The neural correlates of consciousness: an update. *Annals of the New York Academy of Sciences* 1124: 239–261.

Tully, T., Bourtchouladze, R., Scott, R. and Tallman, J. (2003). Targeting the CREB pathway for memory enhancers. *Nature Reviews Drug Discovery* 2: 266–277.

Tulving, E. (1983). *Elements of Episodic Memory*. Oxford: Clarendon Press.

(1985a). How many memory systems are there? *American Psychologist* 40: 385.

(1985b). Memory and consciousness. *Canadian Psychology* 26: 1–12.

(1987). Multiple memory systems and consciousness. *Human Neurobiology* 6: 67–80.

(2000). Concepts of memory. In Tulving and Craik, 33–43.

(2005). Episodic memory and autonoesis: uniquely human. In Terrace and Metcalfe, 3–56.

(2007). Coding and representation: searching for a home in the brain. In Roediger, Dudai and Fitzpatrick, 65–68.

Tulving, E. and Craik, F. (eds.) (2000). *The Oxford Handbook of Memory*. New York: Oxford University Press.

Tulving, E. and Schacter, D. (1990). Priming and human memory systems. *Science* 247: 301–306.

Urban, K. and Gao, W.-J. (2014). Performance enhancement at the cost of potential brain plasticity: neural ramifications of nootropic drugs in the healthy developing brain. *Frontiers in Systems Neuroscience* 8: 38. doi: 10.3389/fnsys.2014.00038.

US Department of Health, Education and Welfare. (1978). *Ethical Principles and Guidelines for the Protection of Human Subjects of Biomedical and Behavioral Research* (Belmont Report). Washington, DC: Government Printing Office.

Valero-Cabre, A., Amengual, J., Stengel, C., Pascual-Leone, A. and Coubard, O. (2017). Transcranial magnetic stimulation in basic and clinical neuroscience: a comprehensive review of fundamental principles and novel insights. *Neuroscience and Biobehavioral Reviews* 83: 381–404.

Van den Hout, M. and Kindt, M. (2003). Repeated checking causes memory distrust. *Behavior Research and Therapy* 41: 301–316.

Van Hoeck, N., Watson, P. and Barbey, A. (2015). Cognitive neuroscience of human counterfactual reasoning. *Frontiers in Human Neuroscience* 9: 420. doi: 10.3389/fnhum.2015.004201.

Van Marle, H. (2015). PTSD as a memory disorder. *European Journal of Psychotraumatology* 6. doi: 10.3402/ejpt.v6_27633.

Velmans, M. and Schneider, S. (eds.). (2007). *The Blackwell Companion to Consciousness*. Malden, MA: Blackwell.

Veselis, R. (2006). The remarkable memory effects of propofol. *British Journal of Anaesthesia* 96: 289–291.

(2015). Memory formation during anaesthesia: plausibility of a neurophysiological basis. *British Journal of Anaesthesia* 115: i13–i19.

(2017). The memory labyrinth: systems, processes and boundaries. In Absalom and Mason, 31–62.

Villain, H., Benkahoul, A., Drougard, A., Lafragette, M., Muzotte, E., Pech, S. et al. (2016). Effects of propranolol, a beta-adrenergic antagonist, on memory consolidation and reconsolidation in mice. *Frontiers in Behavioral Neuroscience* 10: 49. doi: 10.3389/fnbeh.2016.00049.

Volk, L., Bachman, J., Johnson, R., Yu, Y. and Huganir, R. (2013). PKM-ζ is not required for hippocampal synaptic plasticity, learning and memory. *Nature* 493: 420–423.

Volkow, N., Fowler, J., Logan, J., Alexoff, D., Zhu, W., Telang, F. et al. (2009). Effects of modafinil on dopamine and dopamine transporters in the male human brain. *Journal of the American Medical Association* 301: 1148–1154.

Wagatsuma, A., Okuyama, T., Sun, C., Smith, L., Abe, K. and Tonegawa, S. (2018). Locus coeruleus input to hippocampal CA3 drives single-trial learning of a novel context. *Proceedings of the National Academy of Sciences* 115: E310–E316.

Wang, T., Placek, K and Lewis-Peacock, J. (2019). More is less: increased processing of unwanted memories facilitates forgetting. *Journal of Neuroscience* 39: doi: 10.1523/JNEUROSCI.2033-18:2019.

Watrous, A., Tandon, N., Connor, C., Pieters, T. and Ekstrom, A. (2013). Frequency-specific network connectivity increases underlie accurate spatio-temporal memory retrieval. *Nature Neuroscience* 16: 349–356.

Wernicke, C. (1874). *Der aphasische Symtomencomplex, eine psychologische Studie auf anatomischer Basis*. Breslau: M. Cohn und Weigert.

Westbury, C. and Dennett, D. (2000). Mining the past to construct the future: memory and belief as forms of knowledge. In Schacter and Scarry, 11–32.

Wezenberg, E., Verkes, R., Ruigt, G., Hulstijn, W. and Sabbe, B. (2007). Acute effects of the ampakine farampator on memory and information processing in healthy elderly volunteers. *Neuropsychopharmacology* 32: 1272–1283.

Wimber, M., Alink, A., Charest, I., Kriegeskorte, N. and Anderson, M. (2015). Retrieval induces adaptive forgetting of competing memories via cortical pattern suppression. *Nature Neuroscience* 18: 582–589.

Winter, A. (2012). *Memory: Fragments of a Modern History*. Chicago: University of Chicago Press.

Wittmann, M. (2013). The inner sense of time: how the brain creates a representation of duration. *Nature Reviews Neuroscience* 14: 217–223.

Zadra, A., Desautels, A., Petit, D. and Montplaisir, J. (2013). Somnambulism: clinical aspects and pathophysiological hypotheses. *Lancet Neurology* 12: 285–294.

Zannas, A., Provencal, N. and Binder, E. (2015). Epigenetics of posttraumatic stress disorder: current evidence, challenges and future directions. *Biological Psychiatry* 78: 327–335.

Zeman, A., Kapur, N. and Jones-Gotman, M. (2012a). Introduction. In Zeman, Kapur and Jones-Gotman, 2012b, 1–16.

    (eds.). (2012b). *Epilepsy and Memory*. Oxford: Oxford University Press.

Zhou, Y., Won, J., Karlsson, M., Zhou, M., Rogerson, T., Balaji, J. et al. (2009). CREB regulates excitability and the allocation of memory to subsets of neurons in the amygdala. *Nature Neuroscience* 12: 1438–1443.

Index

AA. *See* anesthesia awareness
ACC. *See* anterior cingulate cortex
access consciousness, 87
  precedent autonomy and, 100
actus reus requirements, 171–172
AD. *See* Alzheimer's disease
Adams, Fred, 155–156
Addis, Donna Rose, 20–21
advance directives, 83
advance euthanasia directives (AEDs),
  80–81
agency, 51–60
  with AD, 52–53
  with aphasia, 53
  authenticity and, 136
  in case studies, 59–60
  components of, 51–52
  HSAM, 55–58
  hyperthymesia and, 55–58
  identity and, 128–129
  memory modification and, 126–134
  moral
    through prospective memory, 3
    through working memory, 3
  motor function impairment, 58
  planning and, 53–54
  procedural memory and, 58–59
  prospective memory and, 54–55
  retrospective memory and, 54
  SDAM and, 60
  semantic memory and, 56
  spatial memory and, 33–34
  working memory and, 52
Aizawa, Ken, 155–156
Alzheimer's disease (AD), 48, 78–80
  agency and, 52–53
  personal identity and, 130–131
  prospective memory with, 78–79
  working memory with, 79
amnesia. *See also* induced amnesia
  anterograde, 39, 140

dissociative states and, legal implications
  for, 175–182
episodic, 165
personal identity and, 63–66
  anterograde, 65
  ECT for, 64
  retrograde amnesia, 65
  TGA, 64
  TPA, 64–65
  psychogenic, 140
  as psychogenic, 4
retrograde, 65, 140
  legal implications of, 193–194
semantic, 25
  aphasia and, 39–40, 53
  hippocampal neural prosthetic for, 165
TGA, 140
  legal implications of, 174
  personal identity and, 64
  PPA and, 64
TPA, 64–65
ampakines, 144–145
amygdala
  damage to, 91
  implicit memory and, 108
Andrade, Jackie, 106–107
anesthesia
  amygdala damage and, 91
  anxiety and, 92
  ARAS and, 86
  BIS index measures for, 84–85
  brain monitoring with, 88–89
    with EEG, 88–89
    with PCI, 89
  components of, 86
  conscious sedation and, 91
  consciousness and, 84, 86–91
  goals of, 88
  harm and, 92
  implicit memory and, 106–109
  infused, 89–90

anesthesia (cont.)
  inhaled, 89–90
  intraoperative, 98–111
  memory and, 84, 86–91
  postoperative, 98–111
  preoperative, 98–111
anesthesia awareness (AA), 84–85
  false memories and, 87–88
  GAD and, 90
  PTSD from, 95, 104–105
anterior cingulate cortex (ACC), 33
anterograde amnesia, 39, 65, 140
anti-foundationalism, 22–23
anxiety
  anesthesia and, 92
  GAD, 42
    AA and, 90
  from suffering, 93–94
aphasia, 39–40
  Broca's, 40, 53
  definition of, 39–40
  Wernicke's, 40, 53
ARAS. See ascending reticular activating
    system
Aristotle, 15
  on memory, 15–16
  on recollection, 15–16
ascending reticular activating system
    (ARAS), 86
auditory cues, autobiographical memory
    and, 2
Augustine, 16
authenticity
  agency and, 136
  autonomy and, 134
  memory content disorders and, 113
  memory modification and, 134–138
    cognitive enhancement for, 134–135
    erasure of memories and, 134
  as moral ideal, 134–138
  reconstructive model of memory and,
    113
  as self-discovery, 135
autobiographical memory
  experiential memory and, 4
  hippocampal neural prosthetic as
    influence on, 163–164
  HSAM, 41, 55–58
  personal identity and, 61
  SDAM, 60
  sensory cues for, 2
autonoetic consciousness, 16, 29–30
autonomy
  authenticity and, 134
  precedent, 51, 74–82

  access consciousness and, 100
  critical interests and, 75–78
  defined, 75
  for life-sustaining treatment, 76–77,
    81–82
awareness. See also consciousness; pain
    perception
  harm and, 91–98, 100–101
  memory and, 91–98
  during surgery, 92, 94–98
    memories created from, 97–98
    patient preparation for, 96

Baddeley, Alan, 32
Bartlett, Frederic, 14, 17
benzodiazepines, 98–111
Bernstein, Daniel, 27
beta-adrenergic receptor antagonists,
    116–117
bispectral (BIS) index measures, 84–85
Block, Ned, 87
Borges, Jorge Luis, 56–57, 143–144
brain damage and injury, 42–48
  AD and, 48
  of amygdala, 91
  case study for, 43–48
  of hippocampal complex, 43–48
    personal identity influenced by, 62,
      65–67
  imaging of, 42–43
  memory influenced by, 2
  memory reconsolidation after, 39
  memory retrieval after, 39
  multiple trace theory and, 45
brain function. See also memory systems
  for memory, 1
brainjacking, 159–160
Broca's aphasia, 40, 53

causal theory of constructive memory, 27
CBT. See cognitive-behavioral therapy
Chalmers, David, 154–156
chronic pain, fear memories with, 125–126
Claparede, Edouard, 43–44
Clark, Andy, 154–156
classical conditioning, 35
cognitive enhancement strategies, 142–143
  augmenting in, 142–143
  diminishing in, 143
  for memory modification, 134–135
  optimizing in, 143
cognitive-behavioral therapy (CBT), 126
collective memory, 5
connectedness, 69–70
conscious fear memories, 115

conscious sedation, 91
consciousness
    access, 87
        precedent autonomy and, 100
    anesthesia and, 84, 86–91
    autonoetic, 16, 29–30
    brain monitoring for, 88–89
        with EEG, 88–89
        with PCI, 89
    dissociative states and, legal implications
        for, 175–176
    global neuronal workspace theory of,
        86–87
    neural synchronization theory of, 86–87
    phenomenal, 87
    transition back from unconsciousness, 86
constructive concept of self, 136
constructive model of memory
    episodic memory and, 20–21
    misremembering as part of, 22
    reliability of memory and, 191–192
Cork, Randall, 106
Corkin, Suzanne, 43
cortisol, 103
creative concept of self, 136
critical interests, precedent autonomy and,
    75–78

Damasio, Antonio, 39–40
DARPA. See Defense Advanced Research
    Projects Agency
Davidson, Andrew, 107
Davis, Dena, 79
DBS. See deep brain stimulation
declarative memory. See explicit memory;
    memory systems
deep brain stimulation (DBS), 120–122,
    124–125
Deeprose, Catherine, 106–107
Defense Advanced Research Projects
    Agency (DARPA), 157–158
delirium, postoperative, 99
dementia
    AD, 48, 78–80
        agency and, 52–53
        prospective memory with, 78–79
        working memory with, 79
    early stages of, 74
    euthanasia for, 80–81
        AEDs, 80–81
    genuine change of mind with, 76
    legal implications of, criminal
        responsibility and, 173–174
    Lewy body, 80
    negative rights and, 80

Parkinson's disease and, 80
personal identity and, 130–131
positive rights and, 80
precedent autonomy and, 51, 74–82
    critical interests and, 75–78
    defined, 75
    evaluation of, 74–75
    for life-sustaining treatment, 76–77,
        81–82
SD, 25
Demertzi, Athena, 92–93
Dennett, Daniel, 21
direct realist theory of memory, 4–5
dissociative disorders, as psychogenic, 4
dissociative states, legal implications for,
    170
    amnesia and, 175–182
    cognitive control in, 177
    consciousness and, 175–176
    definition of, 175–176
    somnambulism, 178–182
        imaging of, 181
    testing for, 178
    volitional control in, 177
donepezil, 144
dopamine, 146
Dresser, Rebecca, 81
Dworkin, Ronald, 75. See also precedent
    autonomy

Ebbinghaus, Hermann, on memory,
    16–17
ECT. See electroconvulsive therapy
EEG. See electroencephalography
electroconvulsive therapy (ECT), 64
electroencephalography (EEG), 88–89
Elliott, Carl, 135
emotional memories, 38
encoding, in memory systems, 36–38
engrams, 18
epilepsy, 160
episodic amnesia, 165
episodic memory. See also constructive
    model of memory; memory systems;
    reconstructive model of memory
    altered content of, 29
    autonoetic consciousness and, 16, 29–30
    construction of, 20–21
    definition of, 3, 30–31
    with extended lifespan, 68, 73–74
    in hippocampal complex, 31
    identity through, 3
    legal implications of, 190
    modification of, 127–128
    personal identity and, 61

episodic memory (cont.)
  psychopharmacological interventions for,
    152
  recognition memory, 34–35
    familiarity in, 34–35
  reconsolidation of, 22
  retrieval of, 1, 22
  storage of, 26
  updating of, 51
erasure of memories
  for fear memories, 115–126
    ethics for, 127
    pharmacological interventions in, 127
    selectivity in, 119, 123
  as memory modification, 133–134
    authenticity and, 134
    ethics of, 137–138
    personal identity and, 129
    proxy consent for, 137–138
An Essay Concerning Human Understanding
    (Locke), 16
euthanasia, 80–81
  AEDs, 80–81
experiential memory, 4–5
  autobiographical memory and, 4
  semantic memory and, 4
  trace and, 18–19
  true memories and, 27
explicit memory
  cognitive content of, 30–31
  emotional content of, 30–31
  execution of action plans and, 82–83
  taxonomy of, 15
extended lifespan. *See* lifespan
extinction training, 116

factual memory, taxonomy of, 4
false memories
  AA and, 87–88
  actual experience and, lack of connection
    to, 27
  definition of, 26–27
  implantation of, 28
  NDEs and, 28–29
  neuroimaging studies of, 28
  OBEs and, 28–29
  subjectivity of, 30
  verifiability of, 30
familiarity, 34–35
fear memories, 38, 114–126. *See also*
    memory modification
  beta-adrenergic receptor antagonists and,
    116–117
  CBT for, 126
  with chronic pain, 125–126
  conscious, 115

  erasing of, 115–126
    ethics for, 127
    pharmacological interventions in, 127
    selectivity in, 119, 123
  extinction training and, 116
  hyperactivity of, 126–127
  neuromodulation of, 120–123
    with DBS, 120–122, 124–125
    with ECT, 124–125
    with FUS, 120–122, 124–125
    with TMS, 120
  nonconscious, 115
  panic disorder and, 114
  pathophysiology of, 114–115
  PTSD and, 114–115
    interventions for, 118–119
    xenon for, 118–119
  weakening of, 115–126
    with protein synthesis inhibitors,
      117–118
    with reconsolidation blockade,
      117–118
    with ZIP, 118
fMRI. *See* functional magnetic resonance
    imaging
focused ultrasound (FUS), 120–122,
    124–125
frontal-parietal cortex, 33
functional magnetic resonance imaging
    (fMRI), 42
FUS. *See* focused ultrasound

GAD. *See* generalized anxiety disorder
Gage, Phineas, 130
generalized anxiety disorder (GAD), 42
  AA and, 90
global neuronal workspace theory, 86–87
grid cells, 33–34
gustatory cues, autobiographical memory
    and, 2

harm
  anesthesia and, 92
  awareness and, 91–98, 100–101
  memory and, 91–98
highly superior autobiographical memory
    (HSAM), 41, 55–58
hippocampal neural prosthetic, 156–166
  autobiographical memories and,
    163–164
  brainjacking risks, 159–160
  as closed-loop system, 160–161
  DARPA and, 157–158
  for epilepsy, 160
  for episodic amnesia, 165
  mental time travel and, 163–164

for omissions of memory, legal context
    for, 185–187
as open-loop system, 160–161
recall through, 163–164
for semantic amnesia, 165
testing of, 158
hippocampus
    brain damage and, 43–48
    episodic memory and, 31
    implicit memory and, 108
    personal identity and, 62, 66–67
        case studies for, 65–66
    prospective memory and, 33
    semantic memory and, 31
    short-term memory in, 32–33
    spatial memory and, 33–34
HSAM. *See* highly superior
        autobiographical memory
hyperthymesia, 41
    agency and, 55–58
    HSAM, 41, 55–58
    as memory capacity disorder, 143–144
    personal identity and, 67–68

identity, 50. *See also* personal identity
    agency and, 128–129
    through episodic memory, 3
    with extended lifespan, 68–74
        technological devices and, 71
    in legal implications of memory, 172–173
    memory modification and, 126–134
    in reconstructive model of memory, 24,
        113
implicit memory. *See also* procedural
        memory
    amygdalar mechanisms and, 108
    anesthesia and, 106–109
    execution of action plans and, 82–83
    hippocampal mechanisms and, 108
    priming and, 106–108
    reconstruction of, 107–108
    storage of, 108–109
        in neocortex, 109
    taxonomy of, 15
    as unconscious memory, 106
    updating of, 107–108
induced amnesia, 98–106
    benzodiazepines and, 98–111
    case studies for, 103–104
    consolidation of memory and, 103
    cortisol and, 103
    experimental drugs and, trial studies for,
        105–106
    intraoperative anesthetics, 98–111
    memory-modifying drugs and, 103
    norepinephrine and, 103

postoperative anesthetics, 98–111
postoperative delirium and, 99
preoperative anesthetics, 98–111
propofol and, 98–111
protein synthesis inhibitors and, 103,
    105–106
reconsolidation of memories and,
    102–103
xenon gas and, 99
information technology, 200
infused anesthesia, 89–90
inhaled anesthesia, 89–90

James, William, 17
Jaworska, Agnieszka, 78–80

ketamine, 99
Kihlstrom, John, 106
Kolber, Adam, 188–189

Lacy, Joyce, 191
Laureys, Steven, 92–93
learning, in procedural memory, 1
LeDoux, Joseph, 115
legal implications, of memory. *See also*
        dissociative states
    actus reus requirements, 171–172
    cognitive conditions, 171
        for dissociative states, 177
    criminal responsibility, 171–175,
        193–195
    with dementia, 173–174
    identity in, 172–173
    mens rea requirements, 171–172
    under M'Naghten Rule, 171
    under Model Penal Code, 171, 177
    for negligence, 170
    for omissions of memory, 182–187. *See
        also* episodic memory
        hippocampal neural prosthetic and,
            185–187
        neuroimaging for, 185, 194–195
        reconsolidation of memories with, 187
    personhood and, definition of, 172
    precedent autonomy and, 169–170, 175
    reliability of memory and, 187–193
        alteration of memories, freedom to,
            193
        constructive theory of memory and,
            191–192
        for episodic memory, 190
        for recognition memory, 190–191
    with retrograde amnesia, 193–194
    with TGA, 174
    volitional conditions, 171
        for dissociative states, 177

Lewis, David, 71
Lewy body dementia, 80
lifespan, extended
  connectedness and, 69–70
  episodic memory and, 68, 73–74
  future mental states and, 69, 72
  identity and, 68–74
    technological devices and, 71
  long-term memory with, 72–73
  memory with, 68–74
  Methuselah's paradox and, 71
life-sustaining treatment, precedent
    autonomy for, 76–77, 81–82
Locke, John, 16
  on memory, 16
  on personal identity, 60–61
locus coeruleus, 34
Loftus, Elizabeth, 27, 190
long-term memory, with extended lifespan,
    72–73
Luria, A. R., 56

Madison, Vernon, 171, 173–175
magnetic resonance imaging (MRI), 42
Mandler, George, 15–16
mapping of memory systems, 32
Marquez, Gabriel Garcia, 25, 56
Martino, Pat, 58–59
McDonald, Leslie, 182–186
meaning, construction of, 24–25
memantine, 144
memory. See also agency; legal implications;
    specific types of memory
  anesthesia and, 84, 86–91
  Aristotle on, 15–16
  Augustine on, 16
  Bartlett's reconceptualization of, 14, 17
  brain functions for, 1
  brain injury as influence on, 2
  as collection of systems, 14. See also
    memory systems
  definition of, 17–20
  Ebbinghaus on, 16–17
  engrams and, 18
  ethical dimensions of, 7–8
  with extended lifespan, 68–74
  future of, 196–201
    through information technology, 200
  harm and, 91–98
  historical treatment of, 15–17
  legal dimensions of, 7–8
  Locke on, 16
  neurological disease as influence on, 2
  philosophical theories of, 4
    direct realist theory, 4–5
    representational theory, 4–5
  Plato on, 15

  primary, 17
  processes of, 48–49
  purpose of, 21
  schema theory and, 17
  secondary, 17
  senses of, 19–20
memory capacity disorders, 39, 49. See also
    amnesia; legal implications;
    neurostimulation interventions
  anterograde amnesia, 39
  enhancement strategies
    brain stimulation, 142–143
    cognitive, 142–143
    for memory improvement, 141–144
    for memory restoration, 141–144
  hyperthymesia and, 143–144
  improvement of memory
    through enhancement strategies,
      141–144
    therapeutic interventions for, 141–144
  psychopharmacological interventions for,
    144–147
    ampakines, 144–145
    donepezil, 144
    dopamine, 146
    for episodic memory, 152
    human trials for, 145–146
    memantine, 144
    methylphenidate, 146
  restoration of memory
    through enhancement strategies,
      141–144
    therapeutic interventions for, 141–144
  scope of, 140–141, 166–168
memory content disorders, 41–42. See also
    fear memories; specific disorders
  authenticity and, 113
  etiology of, 112–113
  GAD and, 42
  optogenetics and, 123–124
  personal identity and, 113
  PTSD and, 42
  scope of, 112
memory criterion, 61–62
memory disorders, 4. See also legal
    implications; memory capacity
    disorders; memory content disorders
  research on, 6
memory modification, 126–138
  agency and, 126–134
  authenticity and, 134–138
    cognitive enhancement and, 134–135
    erasure of memories and, 134
  of episodic memories, 127–128
  through erasure of memories, 133–134
    authenticity and, 134
    ethics of, 137–138

personal identity and, 129
proxy consent for, 137–138
identity and, 126–134
as intervention strategy, 139
motivations influenced by, 139
personal identity and, 129–132
first-person subjective aspects of,
130–131
in Gage case study, 130
memory erasure and, 129
third-person objective aspects of,
130–131
protein synthesis inhibitors for, 132
for PTSD, 127–128
of unpleasant memories, 131–132
memory storage
epigenetic mechanisms for, 18
genetic mechanisms for, 18
for implicit memory, 108–109
in neocortex, 109
memory systems and, 36–37
for older memories, 40
short-term, 37–38
neurophysiological mechanisms for, 18
memory systems, 14, 30–42
aphasia and, 39–40
Broca's, 40, 53
definition of, 39–40
Wernicke's, 40, 53
classical conditioning in, 35
consolidation within, 36–38
emotional memories in, 38
encoding in, 36–38
failure of, 40–41
fear memories in, 38
as fixed, 37
hyperthymesia and, 41
HSAM, 41
mapping of, 32
memory capacity disorders and, 39
anterograde amnesia, 39
memory content disorders and, 41–42
GAD and, 42
PTSD and, 42
non-associative learning in, 35
priming in, 35
procedural memory, 35
for prospective memory, 33
ACC and, 33
frontal-parietal cortex and, 33
hippocampal complex and, 33
neural pathways for, 33
for recognition memory, 34–35
familiarity in, 34–35
reconsolidation mechanisms, 36–37
with brain damage, 39
dysfunction in, 39

retrieval mechanisms in, 36–37
with brain damage, 39
dysfunction in, 39
for spatial memory
agency and, 33–34
grid cells and, 33–34
hippocampus and, 33–34
locus coeruleus and, 34
storage as part of, 36–37
for older memories, 40
short-term, 37–38
taxonomy of, 31
memory-modifying drugs, 103
Meno (Plato), 15
mens rea requirements, 171–172
mental time travel, 16, 163–164
Methuselah's paradox, 71
methylphenidate, 146
midazolam, 107
Mill, John Stuart, 189
Milner, Brenda, 43
M'Naghten Rule, 171
Model Penal Code, 171, 177
Molaison, Henry, 66
moral agency
through prospective memory, 3
through working memory, 3
Morse, Stephen, 175–176
Morton, William, 109
Moscovitch, Morris, 17–18
MRI. *See* magnetic resonance imaging
Multiple Trace Theory, 18

Nader, Karim, 37
near-death experiences (NDEs),
28–29
negative rights, 80
neocortex, 109
neural synchronization theory, 86–87
Neurath, Otto, 22–23
neuroethics, scope of, 6–7
neurological disease, memory influenced
by, 2
neuroscience of ethics. *See* neuroethics
neurostimulation interventions, for memory
capacity disorders, 147–166. *See also*
hippocampal neural prosthetic
DBS, 148–151
extended mind theory and, 154–156
criticism of, 155–156
with external devices, 154–156
false memories induced by, 153
in lateral temporal cortex, 149–151
TMS, 147
Nichols, Shaun, 130
non-associative learning, 35
nonconscious fear memories, 115

non-declarative memory. *See* implicit
    memory; memory systems
non-REM (NREM) sleep, 24
norepinephrine, 103
NREM sleep. *See* non-REM sleep

OBEs. *See* out-of-body experiences
obsessive compulsive disorder (OCD),
    35
olfactory cues, autobiographical memory
    and, 2
omissions of memory, legal implications
    for, 182–187. *See also* episodic
    memory
    hippocampal neural prosthetic and,
        185–187
    neuroimaging for, 185, 194–195
    reconsolidation of memories with,
        187
On Memory and Recollection (Aristotle), 15
One Hundred Years of Solitude (Marquez),
    25, 56
optogenetics, 123–124
out-of-body experiences (OBEs), 28–29

pain perception, 92–94. *See also* chronic
    pain
    mediation of, through brain networks,
        92–93
    memory of pain compared to, 93
    suffering and, 93
        anxiety as result of, 93–94
panic disorder, 114
Parfit, Derek, 62
Parkinson's disease, 80
Parsons, Ryan, 117
PCI. *See* perturbational complexity index
Penfield, Wilder, 42–43
personal identity, 60–68
    AD and, 130–131
    amnesia and, 63–66
        anterograde, 65
        ECT for, 64
        retrograde, 65
        TGA, 64
        TPA, 64–65
    autobiographical memories and, 61
    continuity and, 62–63
    dementia and, 130–131
    episodic memories and, 61
    hippocampal damage and, 62, 66–67
        case studies for, 65–66
    hyperthymesia and, 67–68
    Locke on, 60–61
    maintenance of, 63
    memory criterion and, 61–62

memory modification and, 129–132
    first-person subjective aspects of,
        130–131
    in Gage case study, 130
    memory erasure and, 129
    third-person objective aspects of,
        130–131
narrative approach to, 63
Parfit on, 62
TGA and, 64
perturbational complexity index (PCI), 89
PET. *See* positron emission tomography
Phaedo (Plato), 15
Phelps, Elizabeth, 191
phenomenal consciousness, 87
Pitman, Roger, 117–118
planning, 53–54
Plato, 15
Plum, Fred, 86
Posey, Sam, 58
positive rights, 80
positron emission tomography (PET), 42
Posner, Jerome, 86
postoperative anesthetics, 98–111
postoperative delirium, 99
posttraumatic stress disorder (PTSD), 42
    AA from, 95, 104–105
    fear memories and, 114–115
        interventions for, 118–119
        xenon for, 118–119
    memory modification for, 127–128
PPA. *See* primary progressive aphasia
precedent autonomy, 51, 74–82
    access consciousness and, 100
    critical interests and, 75–78
    defined, 75
    legal implications of memory and,
        169–170, 175
    for life-sustaining treatment, 76–77,
        81–82
preoperative anesthetics, 98–111
Price, Jill, 55–58, 143–144
primary memory, 17
primary progressive aphasia (PPA), 64
priming
    implicit memory and, 106–108
    in memory systems, 35
Principles of Psychology (James), 17
procedural memory
    agency and, 58–59
    definition of, 30–31
    learning in, 1
    as memory system, 35
    OCD and, 35
    taxonomy of, 4
propofol, 98–111

prospective memory, 33
    ACC and, 33
    with AD, 78–79
    agency and, 54–55
    frontal-parietal cortex and, 33
    hippocampal complex and, 33
    moral agency through, 3
    neural pathways for, 33
protein synthesis inhibitors, 103, 105–106,
        117–118, 132
proxy consent, 137–138
psychogenic amnesia, 140. *See also*
        dissociative disorders
PTSD. *See* posttraumatic stress disorder

rapid-eye movement (REM) sleep, 24
Rascovsky, Katya, 25
recall, through hippocampal neural
        prosthetic, 163–164
recognition memory, 34–35
    division of, 15–16
    familiarity in, 34–35
    legal implications of, 190–191
recollection, Aristotle on, 15–16
reconsolidation blockade, for fear
        memories, 117–118
reconsolidation mechanisms, 36–37
    with brain damage, 39
    dysfunction in, 39
    legal implications of memory and, 187
reconstructive model of memory
    anti-foundationalism and, 22–23
    authenticity and, 113
    construction of meaning in, 24–25
    episodic memory and, 20–21
    identity and, 24, 113
    implicit memory and, 107–108
    personal identity in, 24
    reproduction model of memory and, 23
    sleep in, 24
REM sleep. *See* rapid-eye movement sleep
representational theory of memory, 4–5
reproduction model of memory, 23
Ressler, Kerry, 117
retrieval mechanisms, 36–37
    with brain damage, 39
    dysfunction in, 39
retrograde amnesia, 65, 140
    legal implications of, 193–194
retrospective memory, agency and, 54
Roskies, Adina, 5–6
Rowlands, Mark, 61
Rugg, Michael, 46

Sacks, Oliver, 59
Schacter, Daniel, 19–20, 190

schema theory, 17
Schulte, Julius, 171
Scoville, William, 43
SD. *See* semantic dementia
SDAM. *See* severely deficient
        autobiographical memory
secondary memory, 17
sedation. *See* conscious sedation
self, concepts of, 136
semantic amnesia, 25
    aphasia and, 39–40
        Broca's, 40, 53
        definition of, 39–40
        Wernicke's, 40, 53
    hippocampal neural prosthetic for, 165
semantic dementia (SD), 25
semantic memory, 1. *See also* constructive
        model of memory; memory systems
    agency and, 56
    definition of, 3
    experiential memory and, 4
    in hippocampal complex, 31
    storage of, 26
severely deficient autobiographical memory
        (SDAM), 60
Shereshevskii, Solomon, 56–58, 143–144
short-term memory, 31–33
    in hippocampal complex, 32–33
    as limited-capacity type of memory, 32
sight. *See* visual cues
sleep disorders, semantic amnesia from,
        25
sleepwalking. *See* somnambulism
smells. *See* olfactory cues
somnambulism, 178–182
    imaging of, 181
Sorabji, Richard, 15
sounds. *See* auditory cues
spatial memory
    agency and, 33–34
    grid cells and, 33–34
    hippocampus and, 33–34
    locus coeruleus and, 34
Squire, Larry, 30–31
Stark, Craig, 191
Stickgold, Robert, 24–25
Strohminger, Nina, 130
suffering, 93
    anxiety as result of, 93–94
surgery, awareness during, 92, 94–98
    memories created from, 97–98
    patient preparation for, 96

taste. *See* gustatory cues
Taylor, Charles, 134–135. *See also*
        authenticity

TGA. *See* transient global amnesia
TMS. *See* transcranial magnetic
     stimulation
TPA. *See* transient psychogenic amnesia
trace, memory and
     definition of, 18–19
     experiential memory and, 18–19
     imaging of, 19
     in Multiple Trace Theory, 18
     Tulving on, 19
transcranial magnetic stimulation (TMS),
     120
transient global amnesia (TGA), 140
     legal implications of, 174
     personal identity and, 64
     PPA and, 64
transient psychogenic amnesia (TPA),
     64–65
true memories
     actual experience as foundation of, 27
     causal theory of constructive memory
        and, 27
     definition of, 26–27
     NDEs and, 28–29
     neuroimaging studies of, 28
     OBEs and, 28–29
     subjectivity of, 30
     verifiability of, 30
Tulving, Endel, 16, 19

unconscious memory, 106
unconsciousness, transition to
     consciousness and, 86
unpleasant memories, 131–132. *See also*
     anxiety; fear memories

VBM. *See* voxel-based morphometry
Veselis, Robert A., 1–2, 108
Vilberg, Kaia, 46
visual cues, autobiographical memory and,
     2
voxel-based morphometry (VBM), 42

Wearing, Clive, 45, 58–59, 66–67, 152
Wernicke's aphasia, 40, 53
Westbury, Chris, 21
The Woman Who Can't Forget (Price),
     55–56
working memory, 1
     AD and, 79
     agency and, 52
     as emergent phenomenon, 32–33
     as limited-capacity type memory, 32
     moral agency through, 3

xenon gas, 99, 118–119

Zadra, Antonio, 178
zeta inhibitory peptide (ZIP), 118